THE GREATEST FUTURE DISCOVERIES OF HUMANITY

THE GREATEST FUTURE DISCOVERIES OF HUMANITY

CLAYTON MURPH

PALMETTO
PUBLISHING
Charleston, SC
www.PalmettoPublishing.com

Paperback ISBN: 979-8-8229-1639-5

TABLE OF CONTENTS

INTRODUCTION

What if the fast radio bursts that researchers have been studying have a distinct purpose? Could they be natural occurrences throughout our universe, or might they signal that we have found evidence of life beyond our very own Earth? Could they possibly be a combination of both the natural and the extraterrestrial? Or maybe they are some sort of space-based GPS system?

Or how about the thought of everything from the past, present, and future coexisting at the same time, an existence where so much more is happening beyond the veil of our reality? Events that have no explanation might be more than meets the eye. Can we truly define what reality is? Could our reality be no more than a program existing on some supercomputer beyond the scope of our reality?

What if the speed of light is actually faster in interstellar space? Maybe our observations do not account for things that could possibly somehow go that speed. I know that from our perspective here on Earth, everything will appear the same. There is a huge chance that interstellar space will change our view of the universe once we get to observe events outside our solar system.

Could a supermassive black hole exploding from the other side have been the cause of our very own big bang? Or maybe something even wilder? The here and the now exist because of this event. Trying to understand how and why may be the greatest question known to researchers. What if the key to understanding this event lies within something that might seem unrelated? Could the very fabric of our universe be explained by looking at things through the looking glass? Could the same big bang have spawned a multiverse existing parallel to our own?

Might the dinosaurs have evolved in a similar way to humans since they had millions of years of evolution throughout three distinct time periods? If we base this theory on how the human species evolved in such a short amount of time, it seems possible that some species of dinosaurs could have easily evolved into intelligent life. These life-forms could have easily been millions of years ahead of humans regarding technology.

Maybe the history of the human race is far more ancient than we have been led to believe, reaching back to a time before modern religion became the explanation for everything regarding our history as a species. What if our history stretches back into antiquity? Or possibly beyond Earth itself? What if the flood happened to destroy a lot of the evidence that would prove that an ancient, technological human species once roamed the earth? What if some of that evidence still exists around the world, ignored by the mainstream researchers because the puzzle is too fragmented to piece together?

Why hasn't the SETI Institute, or any other agency, picked up on signs of life beyond our own planet? So many telescopes of various types pointing out into the universe lead to silence from other forms of intelligent life—if they exist. Despite the absence of signals, we have unidentified craft from beyond Earth that our own government, as well as governments around the world, has acknowledged. This is life-altering for the entire planet because it would suggest that we are not alone after all. The talk of extraterrestrials is not restricted just to the conspiracy theory circles anymore. This is a subject that is being studied by many experts across the globe. Is it time that we find out the truth regarding extraterrestrials?

Let us just think outside the box here for a minute. What if these concepts are not science fiction but science fact? Most people do not take the time to look past our current knowledge. They accept things as they are—answers to questions that we are just beginning to understand. Let me shine a light on things you have probably never even considered. Pushing the envelope always leads to new discoveries. Some of those discoveries have been groundbreaking. As a matter of fact, all discoveries have been groundbreaking, no matter how small they have been.

We would not be where we are as a species if not for people pushing the boundaries in all aspects of life. Some people were ridiculed because they saw things in a different light. They were able to look past the what-ifs to form ideas and theories that have stood the test of time. Some were even put to death for challenging the ruling authorities at the time. The great thing about being alive in this day and age is that we can challenge what is accepted with new theories and new ways of doing things without fear of getting in trouble. In disciplines from mathematics to music, people shaped this world that we see today. Today we hold some of those people in high regard.

As we look back in history, we can see that some people were beyond their time—Galileo, Leonardo da Vinci, Plato, Nikola Tesla, to name a few. They understood concepts that were foreign to their peers at the time. Just because people did not understand those things then, it did not make them any less true. We are living proof of that to this very day. As our knowledge increased, so did our understanding of these very things.

Concepts that were once laughed at are now facts, such as the world not being flat, the earth revolving around the sun, the age of the earth, the universe, and so on. Now we can safely say that some of those concepts are now facts, while others still remain in the realm of theories at the moment. Even now we are still learning new things that we previously did not know. Just because we do not today does not mean we will not know tomorrow.

Some of the concepts I am going to share with you are not widely held by the world. I'm more than confident that as our technology and understanding keep evolving, these theories will either be proven fact or not. As a researcher myself, I know getting it wrong is part of the process of getting it right. As I explain facts as well as my theories on these subjects, you will see things from a new perspective, outside the box if you will. My intentions are that you take this information and push your own thought process beyond your own boundaries. Step out of your comfort zone and see everything with new eyes. Sometimes a new perspective can lead to great discoveries that were basically hiding in plain sight. It is quite possible that you might think of things that I have never even considered. And with that, remember there is no right or wrong answer until it is a proven fact. Imagination is where every idea starts. Without that imagination, our current world would not be possible. We would still be living like our ancestors did thousands of years ago. With all the luxuries we enjoy now, I cannot imagine what my life would have been without them. Our world today shows our progress as a species, the human species.

Our current level of technology is advanced, but it's not at the point of being able to verify things beyond our capabilities. We can speculate all we want through equations and simulations here on Earth. We could be right; we could be wrong. As long as our science is earthbound, we will have Earth answers. I am pretty sure our understanding of physics and the quantum world will change once we venture out into the stars, discovering new elements, particles, and life-forms.

There is a brand-new frontier for the human species to venture out into, beyond the confines of our solar system. An adventure for the first brave humans that leave our solar system behind for exploration into the true, unknown universe. One day in our future, this will become a reality. It's still in the minds of people like me who dream of the day the human race becomes a space-faring race, touching the surfaces of alien worlds for the first time in a display of the progression of mankind.

And inside of this book could very well hold some of the greatest discoveries that will one day be known to mankind. Most of my theories are beyond the scope of our current level of technology, so unfortunately I will not be around to see most of them proven fact or not. Maybe in the distant future my theories could be the basis of something big for the human species as a whole. As long as you take the time to think about them, I have achieved my goal. It never hurts to look outside the box when searching for answers. Outside the box is where innovation lies. Thinking the same way as everyone else will always give you the same results. Pushing the envelope of these discussions is my gift to you: the greatest future discoveries of humanity.

Put your thinking caps on and join me as I journey into the outer limits of possibilities. A place where the wildest ideas could have the answers we have been searching for. Our smartphones, which are powerful handheld computers, were in the realm of pure science fiction not that long ago. Self-driving cars, smart homes, artificial intelligence, Hawking radiation, the Higgs boson particle, sterile surgical instruments, anesthesia for babies, seat belts, and so on—all were wild ideas inside the minds of people who dared to challenge what was accepted as fact. All it takes is that simple idea to change the course of humanity forever.

And the good thing is things are constantly evolving. Look at the evolution of cellular phones, for instance. I remember the brick phones of the '90s; they could not do much more than make calls. I look at my current Android phone, and I can do so much more than just make calls. I can text, watch videos, surf the internet, print, take pictures, play games, live stream, get step-by-step navigation, use wireless signals, and so much more. All this happened since I was a teenager back in the late '90s. I can go on with a lot more things that have drastically changed over the years. The point is that our modern society is the result of ideas and theories of people who have come before.

The ideas held inside this book may very well hold those keys to changing the earth. A new approach will bring a different perspective. A new perspective will result in a different outcome. Together let us examine the facts. Then let's see how these various theories fit the puzzle that we are all trying to piece together.

CHAPTER 1

FAST RADIO BURSTS

Fast radio bursts are something of a real-life mystery. Having been discovered only in 2007 by Duncan Lorimer and his student David Narkevic, they are something new for astronomers, astrophysicists, and amateurs like me. Fast radio bursts are events lasting a fraction of a millisecond to three seconds, releasing as much energy in radio waves as our sun produces in a year. That is truly incredible! Quite frankly, that is a lot of energy. The point of origin of a fast radio burst is one of the last places you would want to be. However, by the time the signal reaches us here on Earth, there is not that much of that energy left. A lot of the punch will have lost its power over the vast distances fast radio bursts usually travel to reach us here on Earth. At the point of origin, you will find the pulse to be very energetic.

Not only does distance dissipate the signal, but time does as well. Remember, every time you look up at the stars, you are looking back in time at how that light looked when it left its origin, however long ago. The starlight we see is light that has traveled over vast distances to reach us. Considering that light travels at a maximum speed of 186,000 miles a second, that is amazing. This equals 5.88 trillion miles a year. This is also referred to as a light-year. Light-years are used to measure distances in space since space itself is so vast.

The fascinating thing about fast radio bursts is that we record hundreds of them a day! That means that there was not a flaw in the original data. After fast radio bursts occurred for many years unnoticed by us people on the ground, this is something big. Most fast radio bursts have interstellar origins in distant galaxies millions and billions of light-years away. Seeing them over that vast a distance is incredible! Being able to trace some of them is some serious astrodetective work. Considering the size of space, tracing fast radio bursts is like trying to find a needle in a haystack. Tracing an event of this magnitude is quite the task. Imagine all the things going on right around us that we have yet to discover. I imagine that some things we have yet to discover are beyond our current level of comprehension. Try explaining a smartphone to an ancient ancestor of yours. Your ancestor would not be able to comprehend what you were trying to say because a smartphone is way beyond anything people understood at that time.

Since the discovery of fast radio bursts in 2007, researchers have learned a fair amount about them. Since these events happen so fast, it is difficult for researchers to trace them back to their points of origin. There are a few that have been traced to distant galaxies' spiral arms, which are the youngest part of a galaxy. I theorize that some fast radio bursts do occur closer to the centers of galaxies as well. Remember, every galaxy has a black hole at the center. So naturally the older stars tend to hang around closer to the black hole than the new stars. The region that is closest to the black hole at the center of a galaxy has more stars in close proximity compared to the stars on the spiral arms. Also, it's good to remember that the spiral arms of a galaxy are where a lot of new star formation is still occurring.

The clue about fast radio bursts originating in the younger regions of galaxies adds other questions. What is so unique about the younger regions of a galaxy that causes fast radio bursts? Why are we not seeing fast radio bursts closer to the center of a galaxy, where the older stars are? Did some of those older stars also experience fast radio bursts when they were younger? I suspect that the answer is yes because things are going to happen regardless of whether we are looking at them or not. Just because you do not observe it does not make it any less true. I did not observe the big bang, yet I know it is a true theory based on a lot of research and work before me.

Through observations, researchers have noticed some odd behavior among fast radio bursts. Fast radio bursts basically fall into two types— ones that repeat and ones that do not repeat. Quite a few repeat on a regular schedule that you can set your clock to. Of course, the schedules of the fast radio bursts that do repeat are not synchronous with each other. However, most fast radio bursts are random events and do not repeat. The fast radio bursts that do repeat are sort of like an interstellar beacon, one might say. Always letting us know that they are here, almost as if purposely.

In all honesty, it is pretty amazing that scientists can detect fast radio bursts over such vast distances in the first place. Most fast radio bursts occur millions and billions of light-years away from us. It's not like they are happening a few light-years away from us. They are happening at distances that are astronomical. Considering the size of the universe itself, this is impressive. Also, taking into account how little we know about the universe in general, I can honestly say we are progressing scientifically as a species. Here is the big question: Are there any fast radio bursts in our own Milky Way galaxy?

Until recently, that answer was no. Then, in 2020, researchers found a fast radio burst in our own Milky Way galaxy that lasted for 1.5 milliseconds. Two radio telescopes, the Survey for Transient Astronomical Radio Emission 2 (STARE2) array, which consists of three radio antennas in California and Utah, as well as the Canadian Hydrogen Intensity Mapping Experiment (CHIME) radio telescope in Okanagan Falls, Canada, detected a fast radio burst named FRB 200428. A very short duration, one that doesn't give a blink of an eye a chance, released as much energy on the radio wave spectrum as our sun does in thirty seconds.

Even though it occurred in the Milky Way galaxy, it was nowhere near Earth. It was traced to a magnetar known as SGR 1935+2154. This magnetar is around thirty thousand light-years from us. Even though it occurred in the Milky Way, it was still pretty far away. On the bright side, this particular fast radio burst was the most luminous fast radio burst ever detected. It was around three thousand times brighter than any other fast radio burst signal to date. I attribute that feat to this fast radio burst being much closer to Earth for us to observe. Researchers were able to see a lot more of the energy at the source. A great deal more was learned about fast radio bursts as a result.

How could the closest one to Earth take so long to be noticed? It was the brightest, so theoretically it should have stood out among the other spatial anomalies. Could it be that more fast radio bursts are happening in the Milky Way and we are just missing them because we are looking elsewhere? The Milky Way is no tiny galaxy either. It is about 52,850 light-years across and holds between one hundred and four hundred billion stars. So there is plenty to miss out on within our galaxy because of the size.

I wanted to include our closest galactic neighbor just to show that galaxies can get to be much larger. The closest galaxy to the Milky Way is the Andromeda galaxy. This galaxy is roughly 110,000 light-years across.

It is also estimated to hold a trillion stars. Compared to the Milky Way, this is a monster galaxy.

Whatever the case may be, I suspect we will begin to see a lot more fast radio bursts originating from within our own Milky Way galaxy now that we have the James Webb Telescope up and running. It will give us a better view of things going on around us because of its unique positioning. Researchers will get a much clearer picture, as well as more accurate data due to the newer instrument onboard.

What could these fast radio bursts mean? What could they be—a natural cosmic event we are beginning to learn about, a sign of extraterrestrials, or how about we meet right there in the middle and say it could be some of both? At this point we really don't know. The leading opinion on where fast radio bursts come from has shifted to magnetars after years of research. Magnetars are mysterious objects themselves, so it makes sense that they are the leading candidates for the origin of fast radio bursts.

In order to fully understand what a magnetar is, we need to hit the rewind button and first take a look at neutron stars. Neutron stars are the collapsed cores of massive supergiant stars that went supernova. I mean stars that are much more massive than our own sun. Our own sun has about one-tenth the mass needed to become a neutron star when it eventually decides to go supernova. Our sun will have a much different fate and end up as a white dwarf after its eventual death. White dwarfs are completely different from the more exotic neutron stars. When the time comes for our sun to end its life in its current form, we will have nothing to worry about. This event won't happen for another six billion years, and none of us will be here to witness it. More than likely the human species will be long gone from the earth and the solar system altogether, quite possibly in a binary star system. This is a discussion for later—back to the magnetars and neutron stars.

The massive supernova explosion causes a gravitational collapse that is so intense that it condenses the core to a diameter of around six miles after ejecting all the material that was the star. A true chaotic event, if you ask me. Neutron stars are the smallest, densest objects known to exist in the universe. After a neutron star is formed, it stops generating heat and begins to cool over time. It really does transform into something completely different. If the star is a lot more massive, the core will collapse into what is known as a black hole. Neutron stars are a step up from black holes in terms of gravity.

Believe it or not, neutron stars can rotate up to a few hundred times a second. Neutron stars are like hyperactive children that cannot sit still. They are in constant motion, reaching mind-boggling speeds. The record for the fastest spinning neutron star goes to PSR J1748-2446ad. This neutron star rotates around 716 times a second, which translates to forty-three thousand rotations per minute. In regular speaking terms, basically a quarter of the speed of light. That is a rotating speed that I find hard to imagine, yet it is true. Watching it rotate at that speed would be incredible to say the least.

Neutron stars are mostly made of neutrons. Neutrons are subatomic particles that do not have an electrical charge—they are completely neutral, hence the name. These small neutron stars still weigh in at about 1.4 solar masses. Small in size but still very massive in terms of weight. A solar mass is around 333,000 times the mass of Earth, to put it into perspective. All that mass packed into a tightly dense ball. Just a matchbox-sized amount of a neutron star would weigh around three billion metric tons! Nothing on earth would be able to lift it.

This is a lot of density compressed into such a small object to say the least. Imagine the intense gravity a neutron star has. A neutron star's gravitational field is about two hundred billion times that of Earth. No way for anything to exist on the surface of a neutron star. Any object would be instantly squashed to thinner than a human hair. Even the strongest materials known to man, such as diamonds and steel, would suffer the same fate. That kind of gravity is very astonishing, yet destructive at the same time.

The only object that has more gravitational pull than a neutron star is a black hole. The gravitational pull of a black hole is around 1.6 trillion times that of Earth. Again, size is the reason black holes formed instead of neutron stars. Either way, both are gravitational beasts that exist throughout our universe. Beasts that are best observed from a safe distance.

Now that I have covered the basics of what a neutron star is, we can dig a little deeper into the mystery of fast radio bursts. The main subsets of neutron stars are magnetars, pulsars, and magnetar pulsars. Out of the three main types of neutron stars, magnetar pulsars are the rarest. So far only a few neutron stars have been found that are both magnetars and pulsars at the same time, which truly makes them rare anomalies.

All pulsars are neutron stars. But not all neutron stars are pulsars. A pulsar is a highly magnetized rotating neutron star that emits beams of radiation out of its magnetic poles. Since they are spinning so fast, naturally pulsars have a wobble to them, similar to a spinning top. These beams, which are emitted out of the poles, are visible to us due to the wobble. The radiation beams can be observed only when a beam is pointing toward Earth, or in the general vicinity. An example of this would be a lighthouse. You see the light flash only when it is pointed in your direction. Of course, you would need to be on the same plane to view the pulse, relatively speaking.

Pulsars pulse at regular intervals that range from milliseconds to seconds. The pulsar is one of the objects in space that sort of chose its own name. As you may have guessed, pulsars were the first candidates to explain fast radio bursts due to their behavior and characteristics. After more observation and research, scientists realized they were on the right track in understanding fast radio bursts, but pulsars were ruled out because they just weren't the right puzzle piece.

This brings us to our next type of neutron star, a magnetar. A magnetar is a type of neutron star that is believed to have an extremely powerful magnetic field. More so than any other type of neutron star. Magnetars also display another characteristic—they rotate the slowest out of any neutron star. Magnetars rotate anywhere from one to ten times per second. This is a lot slower than their fast-spinning pulsar siblings. For some reason, these neutron stars develop a much stronger magnetic field, as well as a slower rotation.

The magnetic field decay powers the emission of high-energy electromagnetic radiation, particularly X-rays and gamma rays. Robert Duncan and Christopher Thompson proposed this theory in 1992. Magnetars are very small objects at about twelve miles in diameter with a solar mass of 1.4. In terms of size, they are the biggest type of neutron star. Magnetars rotate more slowly than other neutron stars, have stronger magnetic fields, and are also the leading candidate for fast radio bursts.

The strength of the magnetic field is what differentiates magnetars from other neutron stars. Magnetars have a shorter lifespan as well. They are a lot more powerful yet have a very short life compared to other neutron stars. After about ten thousand years or so, their strong magnetic fields start to decay. It is estimated that there are around thirty million inactive magnetars in the Milky Way galaxy alone.

A magnetar pulsar is the strangest type of neutron star. To date, only a handful have been found. They have all of the characteristics of a magnetar and a pulsar at the same time. Researchers are having a difficult time trying to understand exactly what is going on here. Are they magnetars transitioning into pulsars or pulsars changing into magnetars? Or these magnetar pulsars could be something else entirely. I theorize that we are looking at something completely new. After more research is conducted, I suspect that they will also show characteristics of other types of exotic stars. From a scientific standpoint, this is exciting because it brings us one step closer to understanding the universe.

In 2020, our understanding of fast radio bursts became clearer. The CHIME radio telescope changed what we thought we knew about fast radio bursts. Like I said earlier, the data points to the spiral arms of distant galaxies as the origin points for these fast radio bursts. Yet there are skeptics who say that if these are the origin of fast radio bursts, we should see fast radio bursts in our own Milky Way galaxy. That is a good point, considering how many fast radio bursts are detected compared to magnetars. The number of magnetars discovered is between thirty and thirty-five depending on the source. Because we have discovered many fast radio bursts, I do not think the answer lies with just magnetars alone—the number of magnetars does not add up to the number of fast radio bursts. What if the explanation is something even stranger than that?

Fast radio bursts originate from fast-spinning objects rotating two to ten times a second with strong magnetic fields wrapping them in a type of shell. Now with everything we currently know about fast radio bursts, current explanations do not give me an answer I am satisfied with. I would like to think that fast radio bursts are something more. At least some of them have to have a purposeful meaning, right?

My first thought was that fast radio bursts could be an alien civilization trying to get our attention. After researchers noticed fast radio bursts, the findings definitely got the attention of a lot of people, including me. I jumped to conclusions hoping for something that it wasn't. After spending years searching, trying to prove that fast radio bursts were signals from aliens, I was disappointed to say the least. I ruled out that fast radio bursts were some way an alien species would try and communicate because the message would not be intact by the time it reached us here on Earth. All the anomalies between the very distant point of origin and Earth would distort and fragment the message to a point where we would never be able to understand it. If fast radio bursts are signals from extraterrestrials, it would be safe to say they are not that smart.

Then I began to wonder specifically about repeating fast radio bursts. Why would a random event repeat at regular intervals? If it is a natural occurrence, what is causing it to happen?

Hypothetically it has been said that a type 2 civilization would be able to harness the total energy output of its sun. Not just by capturing solar energy from the home world but by actually capturing it at the source. For years, I looked at that as the standard. I thought we needed to look for some sort of structure around a star to verify that we are not alone. This is the next step for us humans, right? Wrong! Before I go any further, let me explain the three civilization types.

The types of civilizations are based on the Kardashev scale. Soviet astronomer Nikolai Kardashev proposed this hypothetical model in 1964 regarding energy consumption on a cosmic scale. His scale has three classifications: 1, 2, and 3. Over the years other people have added other hypothetical classifications beyond 3. I have personally stopped after 5 because this scale implied that at that level of technological usage, living beings would have found a way to become godlike. That was the end of that hypothetical rabbit hole for me, so I chose to stick with the standard 3 model.

A type 1 civilization can access all the energy available on its home planet and store it for consumption. This type of civilization may or may not have developed faster-than-light travel. I am pretty sure they would at least have one colony set up in their solar system and quite possibly a few massive space stations to conduct operations in their solar system at least. They would probably be able to mine asteroids and comets that orbit around their host star. From our perspective where we are now, that is still a long way away.

A type 2 civilization can completely harness the energy of a star. Like I theorized, not just any stars—magnetars. This civilization would definitely be a space-faring race. They would have colonies in other star systems in close proximity to their home world. Similar to how things are in *Star Trek*. Very advanced, but still with limitations. I would not mind living during this time period personally. Space would be the final frontier and ripe for exploration! Life itself would be an off-world adventure. But enough about my dreams and back to what I was saying.

Last but not least is the type 3 civilization, which can harness all the energy of an entire galaxy—hypothetically of course. I would assume this would be on the *Star Wars* level of technology, able to destroy an entire planet if they wanted to because of the energy they wield. A full galactic-wide empire with colonies that have evolved into fully thriving member worlds of this empire. It would be very likely that they would have encountered other intelligent life-forms and brought them into this empire as well. Nothing would be out of reach for a hypothetical civilization such as this. The impossible for us would be not only possible for them but also in practice.

As for humans, we are not even on the scale yet. We are still at least two hundred years away from achieving type 1 civilization status in terms of energy consumption. That is saying a lot because we use a lot of energy now as a species. Technology-wise, I would say that we are about five hundred years off, and that is being hopeful. Even though we are not a type 1 civilization, Earth has a very distinct energy signature that another species on our technological level could easily detect.

After coming back to reality many years later, I realized it would not work. For one thing, an advanced civilization building a structure around their own sun to feed the energy needs of their civilization would have a devastating effect on all life on that planet and in that solar system. You would gain an abundance of pure, raw energy while at the same time losing so much more than that type of energy would provide. Would it really be the best way to provide enough energy for a hypothetical type 2 civilization?

Judging from the size of our own sun, I can tell you it would be a very costly endeavor to say the least. The cost would be outrageous. Funding the project would take hundreds of years. Who knows what could happen in the meantime? The money for the project could dry up or a structural disaster could lead to an unfinished project. Just an empty shell floating around your star as a reminder that your people failed on a massive scale.

Finding enough materials would be a challenge as well. It would take mining a good part of your solar system and much more to be able to build something that size. From a human point of view, it does not make sense. It would take at least a thousand years to complete, if not more. The size of that structure would be astronomical. The size of Earth compared to the sun is like a marble to a kickball. How would a species build something that large? It would be larger than our largest planet, Jupiter. Future humans might be able to build modern marvels like that. For now, our current level of technology keeps a project like that in the realm of the hypothetical.

Thinking scientifically, we can see that without the host star, that planet would be cast into darkness. Even the moons would be dark without any starlight to reflect. It would be an icy graveyard for that entire solar system. No way for life to continue on. Not only that, what about the animals and plants that depend on that sunlight? The entire life cycle on that planet would be devastated. That life cycle includes the food chain. I am willing to bet that every species in the universe needs to eat something to survive. Without food, you would starve to death.

Our sun is the reason we exist today. Take that away, and life would have never had a chance on Earth. Earth would be a barren, lifeless planet. The face of the earth would probably be much different. The weather conditions would be a lot more hostile as well. Artificial lighting will never replace the benefits of your host star. The heat the star gives determines whether or not life can survive.

The region around a star where the temperatures are just right for liquid water is the Goldilocks zone—the habitable area around any star. Different-sized stars will have different-sized Goldilocks zones. Some closer to the host star, others further from it. Finding a planet in that region is the tricky part for researchers.

How can you have liquid water without the heat of your host star? Since humans need water to survive, it is hypothesized that all life throughout the universe needs liquid water to survive. In my opinion, that would be suicide for any species because it would send the planet into a cold, dark winter. How would you keep your water liquid? Why waste resources on an artificial source when your own star does a per-fectly good job?

Now a neutron star would make a lot more sense if you were trying to build a hypothetical Dyson sphere. A Dyson sphere is a hypothetical megastructure that surrounds a star for the purpose of energy collection. As we all know, only a fraction of the sun's energy reaches us here on Earth. In order to harness more energy, a structure would need to be built around the star to catch most of that energy. In 1960 Freeman Dyson came up with this theory when he wrote the paper titled "Search for Artificial Stellar Sources of Infra-red Radiation." He theorized that as an advanced civilization increases, so would their energy output.

What type of star is small in size and produces a lot of energy? A neutron star. More specifically a magnetar, due to its immense energy output. It is very small and packs a lot more energy than any other neutron stars. I think any advanced extraterrestrial species would agree with that. It would be a lot more cost effective and manageable. The intense magnetic field would make it a challenge, but not impossible. Remember, you would be dealing with a city-sized object versus something that is much bigger than Jupiter.

Maybe fast radio bursts originate from the hypothetical Dyson spheres. You would not be able to fully enclose the magnetar inside the structure. You would have to leave a couple of openings to release the pressure from the blast of energy or your structure would be blown apart. The energy would need release points at the farthest two points. For a sphere, the farthest points would be the top and the bottom.

Magnetars are the oddballs in the neutron star family because of their characteristics. Since there are so few of them compared to other neutron stars and space phenomena, it is safe to say that something is going on that could indicate the possibility of life beyond our very own solar system.

Building a Dyson sphere around a neutron star makes sense from a logical point of view. If a civilization is at that point to consider building a structure such as a Dyson sphere, they would have already achieved faster-than-light travel. They would be able to travel back and forth to an exotic star that would fit their energy needs. On top of that, you would leave your own host star untouched so life could continue without any hiccups. Plus, you'd have an abundance of energy for your rapidly growing civilization.

What if the materials used to construct the Dyson sphere interact with the energy the neutron star gives off to create this immense gravity field? Somehow this interaction could allow them to slow the star's rotation, which in turn could make it possible for energy to be extracted. This could be a possible explanation as to why researchers have not found that many magnetars compared to other neutron stars. My theory is outside the box, yes? However, my theory is very possible as well.

Any species with the technology to venture out into space is going to need energy to fuel their expansion. Energy powers any civilization, and you need to have a certain amount of energy to sustain that civilization. Like everyone on Earth, we like to get the most we can while spending the least amount. I am pretty sure that is a concept shared across the universe. It allows for other projects that could assist in the expansion of that species. That is, if there is intelligent life in the universe. At the end of the day, I could be wrong.

Another concept I thought about regarding fast radio bursts has to do with a logical question. Again, this theory is outside the box. Hypothetically, if you had the technology to visit any star, any galaxy you wanted, how would you find your way back home? Whatever star you traveled to would have a different night sky, a new view of the stars you would normally see.

Constellations you thought you knew would be gone because you would be looking at everything from a completely different angle. You might not be aware, but most if not all the constellations would disappear. The constellations appear to us in certain shapes and patterns here on Earth because of the arrangement we see. The stars that make up different constellations are not at the same distance from Earth. Some are farther away than the other stars in the same constellation; others are closer to Earth than the other stars in the same constellation. New constellations would dot the night sky. Imagine traveling to visit a distant star on the other side of the galaxy and getting lost. You would not be able to find your way home because now all the stars would look different. What would you do? Send a message home? Even if you could send a message at the speed of light, it would still take years, if not hundreds or thousands of years, to reach back home, so that would not be an option. By the time they would get the message and dispatch a rescue party, you would be long dead.

If you're thinking of simply retracing your steps, it wouldn't work.

Our own solar system is not stationary. In fact, our solar system is traveling around sixty-seven thousand miles per hour, while our galaxy is moving at a whopping 1.3 million miles per hour! Retracing your steps wouldn't do any good because your point of origin would have moved quite a bit by the time you made the return journey, even if you turned right around.

If I were in that situation, there are no words to describe the level of fear and panic I would be in. The thought of being lost in space and never being able to return home is terrifying. Would you give up on trying to reach home? Would you try and find a habitable planet and make the best of things? If it were my choice, I would never give up on trying to reach home. Even if I never made it back, my last breath would be spent trying. I know I said I want to travel the stars. But I meant with the ability to return home in my lifetime. In a constantly moving universe, you would need something that stood out above everything else as a navigational guide, or you would end up lost in space.

Again, looking at the origin point of those stationary fast radio bursts, what if these fast radio bursts are part of a massive GPS system used by some highly advanced extraterrestrial species to navigate the universe? You would need something to emit a massive amount of energy to be detected billions of light-years away. At this present time, only one object fits that description: magnetars. Imagine an advanced species manipulating magnetars in such a way that they produce fast radio bursts. All those random fast radio bursts detected throughout the universe could be ships pinging the large relay stations, which are stationary fast radio bursts, to use them as navigational guides.

When you think about it, it makes sense. If you had a stationary one by your home planet, you could always find your way home regardless of where you were in the universe. Since fast radio bursts are so bright and energetic, why wouldn't they be used as a space navigational system? If a species could travel freely through their own host galaxy, what's to stop them from going to another galaxy for exploration? Or even beyond that to a place that is beyond our current level of comprehension?

To further add evidence to my theory is that until April 2020, there were no fast radio bursts known to have come from our own Milky Way galaxy. Now we have spotted one in our own backyard. It was one of those random fast radio bursts that researchers normally pick up on all the time from outside the Milky Way. If you were to take my theory into consideration, it could very well be an expanding extraterrestrial species that is exploring this region of the universe, an empire stretching out across the night sky of a distant galaxy all the way to the Milky Way galaxy. If that is the case, it will really make us think about our place in the universe.

That could easily be a hypothetical type 3 civilization. A civilization that is way beyond our science-fictional fantasies and imagination. Hypothetically speaking, could a civilization exist that is basically on the level of *Star Wars*?

Researchers have been picking up fast radio bursts from other galaxies. Then boom! One pops up in the Milky Way. Fast radio bursts are detected from as far away as billions of light-years. Many come from millions of light-years away. That is an enormous distance to travel for our detection. Way beyond the scope of what the human eye can achieve. And way beyond anything we can come up with at this present time.

Just thinking about why we have never detected any fast radio bursts from our own Milky Way galaxy until now brings a lot of questions. As far-fetched as it may sound, what if these fast radio bursts could lead us to the discovery of other intelligent life in our universe? I am curious to see if there is a way to track the spread of fast radio bursts, like a point of origin from when and where fast radio bursts started to where they are currently happening. We probably will not be able to do that because we are extremely late to the party, but it is a thought. But considering the fact that one fast radio burst has been observed in the Milky Way, this could be the frontier region of a hypothetical alien species. And maybe they are just now starting to map this region of the universe where the Milky Way is located. This is just a theory, of course.

The dark side could be that the fast radio burst detected in our Milky Way galaxy is the result of an expanding species. Our worst nightmares could come true. Nobody willfully gives up territory unless they are on the losing side. The late Stephen Hawking warned us about trying to contact an alien civilization because he referenced how badly it went for the Native Americans. Honestly, I couldn't agree more. If a hypothetical extraterrestrial civilization arrives in Earth's orbit and decides to wage war against humans, we will lose this war in a matter of hours, days at the latest. Let's not think that first contact will happen in such a tragic manner. It is very likely it will be a great peaceful event that will uplift the entire world.

Wondering if only this is true? A wild imagination can sometimes be closer to the truth than we would like to admit. Until the day we can go investigate, or Earth gets invaded, your guess is as good as mine. But I would like to believe that everything has a purpose and a reason. Though we might not understand it at the time, a reason is always there. Same thing with fast radio bursts. They might be naturally occurring phenomena in our universe. Or they might be an instrument for use by another life-form. Until we have solid evidence one way or the other, there is no right or wrong answer. My own theory is that fast radio bursts are part of a space-based GPS system. The how, the who, and the why are the big questions. That is a discussion for another time that I hope to be having with you.

CHAPTER 2

THE BOOK UNIVERSE THEORY

Several years back I was reading stories about the ghosts that have been seen on the Gettysburg battlefield throughout the years. I was trying to find some kind of explanation for ghosts because there really isn't one. A lot of people, including me, have seen ghosts. Before my own experience, I doubted the whole idea of ghosts. I truly believed it was people faking these stories for publicity. Some stories were proven to be hoaxes. Some people even admitted to deceiving others because it would make them money. I was very skeptical when it came to ghost stories. That is, until I saw one myself.

Back in the summer of 2018, I was walking down the hallway of my house, headed to my bedroom. As I passed my daughter's room, I glanced in and thought I saw her playing with her toys in the middle of the room. Immediately after passing her room, I turned around to go check on her because she seemed very lonely. It wasn't even a second later when I looked back in her room, and it was empty. So I went in and looked around, thinking she was trying to hide and scare me. I searched everywhere in her room; she was gone.

I went into the living room, where my other children were. I saw my daughter on the couch watching television with her brothers. I asked her how she got out of her room so fast without me seeing her. My children gave me a very puzzled look like they had no clue what I was talking about. All five of my children told me she had been watching television with them the whole time. They were very serious when they were telling me this. For a minute, I thought I was going crazy!

After dancing around the question for several minutes, I finally told my children what I had seen in her room while passing by. My daughter said she had also seen this little girl in her room several times, laughing and playing. What my daughter described is exactly what I saw.

Over the years, I found out my daughter and I were not the only ones to see this little girl. My two oldest boys, who are in their late teens, told me that on more than one occasion they have seen a little girl about the same age as their little sister walking in the hallway at night long after everyone is sleeping. One son told me he used to go check in the hallway to see what his sister was doing, only to find the hall empty. He said he would go into her room, and she would be in a deep sleep. He stopped checking not too long after that.

Last fall my mother-in-law came to visit for a month. She shared the room with my daughter since there was a second bed in there. Almost every morning she would complain about my daughter sleeping in the bed with her at night. My daughter denied it because that is not something she would do. My wife suggested our daughter stay in our room for a few nights to see if that would please her mother. Turns out it didn't. Then she was accusing my wife of lying with her just to mess with her.

We all tried telling her we have a houseguest who stays in that room as well. Mother-in-law did not believe a word we told her. She does not believe in the supernatural or anything that is not already explained and verified. At this point, I just wanted a rational explanation of what was going on in my house.

I knew I had to try and figure out what was happening because what I experienced defied logic. I was never able to find out anything about a strange death or murder at my place of residence. I had no clues to go on. I was stuck with no end in sight. I kept on thinking about my experience as well as other people's experiences.

My experience made me rethink my beliefs about ghosts and other things of that nature. I explored many theories over the years but came up empty-handed on all of them. I was almost at the point of giving up on trying to find an explanation for ghosts when, by chance, after watching a *Doctor Who* episode, something clicked.

I remember thinking that maybe those events and the participants of those events became fixed in a point in time, doomed to repeat the same set of events forever and ever. Basically a time loop. But this theory does not work since time is linear and flows in one direction. Time does not loop! There is no way to hit a rewind button and live the day over again. Unless time is not what we think it is. I guess only time will tell.

Yet somehow the energy of people from the past is present from time to time. How can that be unless things are not what they seem?

What is the connection I am missing? For many years I tried to tie everything together, but how? After years down this path, another question came to mind—what if past, present, and future exist simultaneously? What if the only time that exists is the present? The here and now and that is it? Basically, every moment in history and the future is happening right now. Even while you read this, the events of time are going on in our present. Dinosaurs are running around beside us, but we would never know.

This theory would change the way we view time. Just the thought of that being true would explain many things. Could it be the explanation for the Mandela Effect as well as other strange things that people have reported over the years?

Before I go any further I'd like to give you an insight into what the Mandela Effect is. If Nelson Mandela came to mind, then you're on the right track. Even though Nelson Mandela died in 2013, many of us remember him dying in prison in the 1980s. In 2009 the term "Mandela Effect" was born. The more people started thinking about it, the more things came to mind, such as widely misremembered spellings of names like Jif peanut butter, the Berenstain Bears, Looney Toons, *Sex and the City*, Febreze, and Froot Loops, just to name a few. Everything I just mentioned involves false memories that are shared by many people, including me. I could seriously go on and on about this because there are so many instances. I hope this answers any questions you might have had before we went further.

Instead of time being linear, I theorize that time is, in fact, stacked—similar to sheets of paper in a book. When you read a book, you read every page the same way regardless of what page you are on. As you read each page, you get further into the book, further into the story. Each day we experience in the present gets cataloged away in this book that makes up our current reality.

The binding connecting every page of that book would be something that we would be able to interact with in some shape, way, or form. What if time itself is that binding, linking each and every event that we call the past, present, and future as a sheet? Hypothetically, since time itself is that binding, every page in the book should already exist. Not only that, but we should also be able to access any page in the book since we are a part of the greater story, possibly with time travel or interdimensional travel, depending on which theory you prefer. I will talk about that in another chapter.

Our universe is a strange place that goes beyond our current understanding of physics. As our technology advances to the point where you cannot tell reality from computer-generated images, it makes me wonder what reality is. Look at superhero movies and sci-fi flicks. They look very real to me and a lot of other people. Even video games are so realistic these days. In a few years, I imagine video game graphics will be able to mimic reality to the point where it would be hard to distinguish between the two. Some people will genuinely be confused as to what is real and what is a digital creation. How would a person know the difference? The truth is that a person would not be able to tell.

As we all know, video games are created from the start to the finish. When you first get a game and play, you do not know the outcome of the game until after the first time you play it all the way through. Most video games have a story mode that takes you on a journey. You get to experience the life of that character over the course of that game. You live parts of his life, so to speak.

As with any game, once you complete a level, you can always go back and replay that level whenever you want. If our reality, our universe, is in fact a massive simulation, that may indicate that there is an endpoint. Not only that, but everything is already there, layered in the way that the story is told. The entire story already exists!

It was hard to comprehend at first. After research, it became clear to me that our reality is a lot more complicated than I imagined. You are probably wondering how the future is already determined when current events change the course of the future. How do we know what that future was supposed to be until we experience it? In reality, how do we know if current events change the future? We cannot see the next page in the book, so we really don't know. From our point in time, anything can happen, right?

If you look throughout history, you will find certain individuals who could see other pages in this book—these individuals were known as prophets. They were able to see things that were yet to come. How could a prophet make a prophecy about future events unless those events already existed? Those events will happen regardless of what we do because they were meant to happen.

My theory is that the future is already determined because it was already written. I know it sounds like a religious concept because those words are also in the Bible, but I am not going to discuss religion with you. My intentions are to go another way.

Let me ask you this: What is consciousness? Yes, we are alive and self-aware, but still, what is consciousness? A new theory out of Boston University's Chobanian & Avedisian School of Medicine has really caught my attention because I believe it's a major piece of this puzzle. The basics of this theory are that we do not perceive the world, make decisions, or perform actions directly. Instead, we do these things unconsciously, and then, about a half second later, we consciously remember doing them, according to Andrew Budson, medical doctor and professor of neurology. When I read this, I thought this new theory was huge, because it could also suggest that we are unconsciously acting out hypothetical programming.

This new theory takes choice away from us. The choice was already made before we even realized it. I remember many years ago thinking about this exact subject. I know I can reach out and grab the glass of water in front of me, but why? How? I understand the choice, but not how it came to be. How do I make the choice? It's almost as if my decisions are already predetermined.

With this new theory of consciousness, it appears that this very well could be the case. If our actions are predetermined, that means that there is no free will. Choice is nothing but an illusion. This could also imply that our lives are not as real as we thought they were. Our entire reality is not what it seems. Then the question shifts to this: Who or what is making the choices for us? To me, this is a major step toward proving we are living in a massive simulation.

It might also be a rational explanation as to why some people make bad choices, choices that everyone else knows are wrong. A higher form might be guiding those choices the same way people playing *Grand Theft Auto* make those same kinds of choices.

When I first heard about the simulation theory, I dismissed it as pure science fiction. I thought there was no way possible for people to exist in a simulation, so it must not be true. Then again, twenty years ago I never thought it was possible to have a cellular phone that is a computer either. So of course, my opinions evolved over the years as our technology advanced. I see things now that I never thought possible years ago. Around that same time, I realized I could be wrong about the simulation theory.

Believe it or not, the book universe theory fits within the framework of an advanced simulation. When you study the concept and examine everything around you, you will start to notice things that you previously didn't. Why can the speed of light not be broken? Is it the ultimate processing speed of the computer running our simulation? Can nothing go any faster because the massive quantum computer is at its limit in terms of processing speed?

The precise mathematics is abundant in nature, almost as if everything was designed perfectly. The golden ratio appears everywhere in nature. From flowers to shells and galaxies—it is there in plain view. Why is our universe so mathematical in nature? This is a big clue that screams simulation! The odds of our universe being here are very low. I am talking about one in 102,685,000, yet here we are. This is a lot more than coincidence!

Everything needed for complex life-forms to evolve into what we see today could not be merely chance. Everything has to be designed by an intelligence that is far beyond the human mind and capabilities. Why would an advanced species need to create a simulation so detailed if not for a specific purpose? Are we being observed directly? Or are we some secondary characters in this simulation that do not garner any attention from the creators? Your guess is as good as mine. It really does open up more questions than can be answered at this time.

My biggest question regarding simulation theory is this: Who are they? Who or what could have created this massive simulation? Scientists run computer simulations all the time trying to test theories and disprove others. The simulations they run would be considered sticks and stones compared to the one we hypothetically exist in. It really isn't the same.

Another thing I wonder about regarding the simulation theory is time. Hypothetically, if we are existing inside of a simulation, I wonder if the way we experience time is different from the way time works outside the simulation. If you have ever played one of those simulation games where years end up passing in the game but you have been playing only an hour or so, then you'll notice the character you are playing will have aged quite a bit during that time, while you (the player) will look exactly the same as when you first started playing the current game.

What if our simulation experiences time in this rushed fashion? It might be the reason why time is experienced differently by everyone. In theory, our entire universe, our existence, could be viewed in a matter of hours if in fact we are inside a massive simulation.

Even when you add religion into the mix, it makes you wonder. According to the Bible, everything is already written. That hints to the possibility of everything being predetermined. And science itself is pointing to our very existence being inside a simulation. Again, the parameters and outcome of the simulation are already predetermined. Looking at things from this point of view would imply that choice is just an illusion. We are basically going through the motions of our programming, if you will.

The bigger question is this: Why was this simulation created? By chance, what if we are the creators of our own simulation? When I say *we*, I do not mean right now in our lifetimes. I am referring to humans. More specifically, humans from the far distant future. I theorize at least fifty thousand years, if not more.

The best way to describe the simulation would be sort of like *The Matrix* in a way, just on a much bigger scale. I use *The Matrix* as a reference because everyone existed inside a shared computer program. Our shared reality could very well be similar in many ways to the world of *The Matrix*.

Hypothetically, why would humans create a simulation? Maybe the purpose of this simulation is to experience life the way things used to be? All the way from the very beginning to the very end. What if the universe already ended in its current form? All planets hidden in eternal darkness, frozen over in a brutal cold after a trillion years of expansion pushed all the galaxies so far apart that they are beyond detection by any means. Galaxies eventually lose shape, and stars drift apart. Eventually every star will die in its own way. The night sky would be completely nonexistent. Nothing left but endless darkness.

What if the entire human race is traveling on a planet-sized ship through an endless, empty space for eternity? What if all humans are somehow plugged into this simulation while their bodies lie in a form of cryogenic sleep? Could it be that humans are in fact eternal outside this simulation? Unable to die, nothing left to see or do—the only thing left is to relive the human existence. If so, it could be a possible answer for reincarnation.

What if the only thing left that exists is the record of the human civilization until it was no more? Basically, humans have already reached the very end of everything. We relive the story of man, over and over. That would be a reason to run a massive simulation. And since we are basically reliving past events, every event has already been cataloged into the book. The future was already predetermined because these events happened outside the simulation in the so-called real world. The story is always the same; it never changes because it is already documented history. So the events that we experience in the here and now are nothing but a mere page in the book of life. It really gives a new meaning to "it was already written" from a nonreligious standpoint.

The hole goes deeper, obviously. But how deep is what I want to know. Looking at how the book universe theory fits within the framework of the simulation theory and aspects of religion at the same time is exciting! It is like they are different parts of the same thing, trying to tell the same story from different points of view.

About ten years ago, I got a Nintendo Wii for my young children at the time. I remember going through the whole "create your Mii avatar" process. My wife and five kids and I made a total of seven avatars. The first thing we noticed was that even though we'd just designed our characters, ours had so many similarities to a lot of other avatars people had already created. There were a lot of copies with slight differences.

Furthermore, in terms of the Mii avatars, there is only a certain number of avatars you can make variations on. Eventually those variations will begin to overlap many times over. It dawned on me. What if humans are created the same way within this simulation? That would truly explain the doppelgänger effect. I've always wondered why so many people look similar but aren't related.

It was only after that I began to think about people. I have lived in four different states and visited others. I always see people that I think could be related to others I know from a previous place because of the almost identical appearance. Even when you look at some historical photographs, you will see similarities to some people who are alive today. When you look at the big picture, you can see that it is more than a coincidence.

Having moved around a lot, I know people always stop my wife and ask her if she is so-and-so because "You look just like her." The first couple of times it happened, I thought nothing of it. I thought it was strange, nothing serious. After hearing it over and over, I started to wonder why.

The last incident I recall was in late 2019, right before COVID-19 captured the attention of the world by its rapid spread. At the time, my wife was working at a hospice care center. There was an elderly gentleman who kept calling my wife another name, the name of his daughter. He would always tell her she looked exactly like his daughter. She always brushed it off because she thought he was just senile.

When that man's family came, they all made the same mistake and mistook my wife for their relative. Hugs and kisses, I mean the whole thing! They even showed her a picture, and according to my wife, they looked like twins. My wife explained to them that she is from California, has Mexican heritage, and speaks Spanish fluently, and her paternal grandparents still live in southern Mexico. Only then did they realize they'd made a mistake. She was not Native American like them.

How can that be? How can my wife have a double when she grew up on the West Coast and this happened in rural southwestern Virginia? My mind was everywhere, trying to find a solution for what my wife was telling me. At first, I did not know how to proceed. After I did some research, I realized this was a doppelgänger event.

People find their exact doubles all the time. Some even have the same types of personalities, hobbies, and interests. Some have the exact same looks; some have a slight variation such as height, weight, or body shape. It's obvious there is a pattern.

New findings indicate that doppelgängers share DNA as well! Even though doppelgängers have no family relations, they are still connected at the microscopic level. In theory, you could loosely use the word *copy*. But how can that be? Who were the originals we were modeled from?

Even looking over old photos, you will find resemblances among the living, like a pattern with clues left in our visual appearance. I don't think this is a coincidence either. The clues were left for us to follow for a reason. So in a sense, there is always some version of us on the earth at any given time. Hypothetically, this could be seen as some sort of reincarnation.

Like I said, I am not here to discuss matters of religion. I am referencing different concepts to show the overlap regarding certain things and only highlighting a connection that can go much deeper if fully explored.

Now that we have explored the simulation theory, let's take the conversation in another direction and examine other things that also tie into the greater theory. The simulation is the program; my book universe theory is the parameters of the program. What if the ghost stories we have all heard, and some of us have experienced, have a much more rational explanation? Ghosts could be nothing more than one page bleeding onto another in a sense? Even though the event already happened in our past, we could be getting a glimpse of that past in real time. We could be seeing an event actually happen, whether it be just a simple walk across the room, the playing of the piano, or cleaning, for that matter—seeing it as if we were there in basically a time-locked event.

It makes me wonder if they can see us as well. I mean, if we can see them, I suspect they can too. If a person is seeing a past moment, do the people of that moment see the present person in the same way we would see a ghost? Would we appear as a see-through object to them as well? Would they also see our technological items? Would we be aware that we were existing in both time periods at the same time? A question I'll unfortunately never know the answer to.

Hypothetically, what if you were unaware that you were somehow existing in both time periods? Or what if, for that brief moment, you actually left your own time period and existed only in that out-of-place time period until you returned to your normal time? Perhaps something else entirely?

As it stands, these events happen only in certain areas, some traumatic, other areas not. Quite possibly a lot more areas can have this overlapping effect if you look at this from a logical standpoint and realize that every inch of land on the earth has some form of traumatic history attached to it. It is more than likely that every area has its own story that goes ignored by most of us, me included. Maybe people are too busy in their everyday lives or distracted with their phones to notice?

At the present time, there is no concrete explanation as to why some spots have paranormal activity while others do not. Could it be that the binding is weak in those areas after having existed for billions of years?

Maybe those are areas that could allow us access to other pages? Could these types of events be widespread throughout the entire universe? Other areas in the universe might have places where two time periods exist at once in a chaotic mess, like a merger of both sides, if you will. A sight to see, but from a safe distance. Again, this is just a theory of mine.

The only way to find that answer is to travel to other places and see for ourselves. But I am pretty sure we are not going to be thinking about ghost hunting when we do eventually venture out into space. I suspect that future astronauts will find things that will be a lot stranger than ghosts. Things we have never seen before or thought possible. Unfortunately, it will be a very long time before we reach that level of technology as a species to achieve faster-than-light travel. Hopefully we will reach that technological level that will allow us to investigate different planets one day, perhaps.

I have always believed in the possibility of extraterrestrials. Out of all the galaxies and stars, we cannot be the only life in the universe. What if my beliefs are misguided? Not that extraterrestrials don't exist, but perhaps some unidentified craft could be from a different time period? The unidentified flying objects that have been reported over the years could very well be future humans going on about a regular day. In a sense, ghosts from the future? Not aliens, but humans! Humans from far off in the future.

This could quite possibly be the reason unidentified flying objects haven't stopped to say hello. If these crafts are piloted by future human beings, they would not even realize they are being seen by people from the past. Since we do not know what's on the next page, anything beyond us is stunning and unidentified.

There could be different levels to the technology that has been reported over the years. There have been reports of several different-looking ships among the government and UFO circles. What if each different type of ship corresponds to a separate future era? Like the pyramid-shaped ones are from a hundred years into the future, the triangle ones a thousand, and the classic disc-shaped UFOs a hundred thousand years into our future?

This would sure give the UFO phenomenon a flip upside down. Not visitors from a distant star system, but rather humans from the earth of our future. A time when our science fiction is actually science fact. To us, any ship beyond our current capabilities is basically classified in the same category—unidentified flying object. We have grouped numerous crafts into this category. Even the unidentified submersible objects are a subject of this category.

What if future humans have figured out how to access the binding of our book? In basic terms, time travel. And what if those unidentified flying objects that are always reported are actually time machines? Things we see by accident on our current page could be the norm for our future selves. Could it be that UFOs are actually time-travel vehicles piloted by humans?

If all this is true, everything we thought we knew about time and the occurrences in it would change. Everything that is ever going to happen, everything that has already happened, along with our current lives, could be happening right now in the same shared space. Might it also be the answer to dark matter? What if dark matter is only the veil shielding us from other pages in the book of time? We see only a fraction of what is in space because dark matter makes up the rest of it. We are still trying to understand and define what dark matter is. Dark matter could very well be the space of that time, the sheet of paper, as it were. I wish I had a definitive answer for you.

I believe we should go even deeper down the rabbit hole to another theory that fits within the simulation/book universe theory, and that is the multiverse theory. More and more scientists are open to the idea of the multiverse theory. Theoretically our universe could be just one out of an infinite number of universes. Hypothetically, each universe could be different from the next. Some might not have any heavy elements. Others might just be exotic particles.

Maybe each universe is the same, and the only difference is the choices that were made. Instead of turning left, you turned right or went straight. You might have stayed home or caught the bus. Maybe somewhere out there each choice exists, every possible outcome to every situation. With a simulation you would have multiple outcomes depending on certain conditions that could affect the choice, similar to the conditions under which scientists use mice. We would be considered the mice if you look at it that way.

Looking even deeper, we can postulate that some version of us exists in every universe. Each person is just a sliver of the whole. That brings a lot more questions and possibilities than I can focus on regarding the whole theory. Over time it might be easier to understand the multiverse aspect of the theory.

The book universe theory is basically the framework and parameters for this simulation, sort of like the cover and binding on a book—hence, the book universe theory. This is one topic I suggest you research further. I myself started my research with Nick Bostrom from Oxford University. From there, there were many avenues to take in my hunt for the answer. I believe the further outside the box we go, the clearer the picture will become. As we make advances in quantum mechanics, we will finally be able to see this framework in real time. Over time we will eventually have an understanding of what we are looking at. Our place and our purpose in this book will one day be revealed. Until that day, don't stop searching.

CHAPTER 3

THE SPEED OF LIGHT

One hundred eighty-six thousand miles per second is an incredible speed to say the least. In fact, that is the maximum speed limit for anything in our universe. No matter how you measure it, the speed of light is always the same; it never changes. Albert Einstein's theory of relativity gave us the equations we use to determine this. Our current understanding of physics says no object with mass can ever reach the speed of light. If an object traveled at the speed of light, its mass would increase exponentially. At that speed, its mass would become infinite, and you would need an infinite amount of energy to move the object, which is more energy than our entire universe can ever produce over its entire lifetime.

Also, as you got closer to light speed, you would experience time dilation. Time would slow down for the person traveling at 90 percent the speed of light. You would probably experience serious tunnel vision, I theorize. At the same time, we have never gone that fast, not even close. Our fastest speeds would be considered a slow crawl compared to the speed of light.

The fastest object we have is the International Space Station, which orbits Earth every ninety minutes at around 17,500 miles per hour. That is blazing fast for anyone alive. Still, it is not fast enough to visit another solar system in a lifetime. So if you're planning a trip to Qo'noS anytime soon, I'd reconsider.

Our understanding of light comes from observations right here on Earth. That does not really give us much to work with when you think about it. But what if we are wrong about the speed of light? I know it sounds absurd, but what if? I am no physicist. I'm just a curious human being who is fascinated with every aspect of space.

In the vast universe, light may be the king of the speed game. Or maybe objects that travel faster than light cannot be picked up by our instruments. Maybe going faster than light speed, you slip into a layer of subspace that is beyond our detection. What if the speed of light appears to be constant to us on Earth because of our location?

Our solar system is enclosed within a bubble of charged particles that originate from the sun, called the heliosphere. The heliosphere is the region of space surrounding the sun as well as the planets. This region extends to the boundaries of our solar system. These charged particles produce a solar wind with an outward direction that acts as a shield, protecting us from the harmful radiation that is beyond our solar system.

Within this heliosphere, multiple things are happening at the same time. For instance, the solar wind is slowed and diverted by the interstellar medium, which gives the heliosphere its bubble-like shape. Not only that, the magnetic field of our sun is also deflecting and directing particles.

This bubble is around 100 AU from Earth depending on the direction you measure it by. An astronomical unit, or AU, is the distance from the sun to Earth. Scientists use this unit of measurement to calculate the distance of objects within our solar system. Outside our solar system, distances are measured in light-years. A light-year is determined by the distance that light travels in a vacuum in one year. That number is about 5.88 trillion miles a year.

To illustrate the magnitude of trillions, I'll use seconds. A million seconds is twelve days. A billion seconds is thirty-one years. A trillion seconds is 31,688 years. Regarding time, modern civilization hasn't even existed for a trillion seconds. Distance-wise, the nearest known star to Earth is Proxima Centauri, around four light-years from Earth, or twenty-four trillion miles. So the nearest star to us, if we were traveling at the speed of light, would still take us four earth years to get to. Where's Zefram Cochrane when you need him?

However, there is a theory on warp drive engines thanks to theoretical physicist Dr. Miguel Alcubierre. The concept is that you would not travel faster than light. As a matter of fact, your ship would not move at all. You are probably scratching your head right now. Don't worry, I will get you up to speed—warp speed, that is.

Light may have a speed limit, but the fabric of the universe does not. Space itself is expanding faster than light. When the big bang occurred, our universe exploded into existence and spread out across vast distances in a few fractions of a second. Obviously, the speed of light did not exist at the very beginning because there were no stars. Star formation didn't take place for another couple of hundred million years after the big bang.

Now back to what I was saying. Hypothetically, this warp drive engine would allow us to contract space in front of the ship and expand space behind the ship, resulting in faster-than-light travel. In theory, all this could be achieved without breaking the laws of physics—shifting the space around an object so that the object would arrive at its destination before light would in normal space.

So on paper we can effectively travel faster than light without breaking the speed of light. Confusing? Yes, it is. Physics is its own language. One thing I am discovering on my own is that our understanding of things evolves each and every day. What we think we know today can always change tomorrow. The beauty of it all is that you keep learning. If we stopped learning, we would still have the intelligence of cavemen, if you don't mind my saying.

Remember I was saying our solar system resides inside a bubble called the heliosphere? What if our observations of light speed are distorted because of this bubble? Let me explain. Let's say the heliosphere acts like the boundaries of a city. Inside the city limits, you have speeds that are much lower than those on the open highway. Take a look at our own interstate system as a reference. On the open roads, the speed limit is a good seventy-five miles per hour. As soon as you enter into the city limits, the speed limit decreases to fifty-five or sixty miles per hour. Once you exit the city, your foot goes back on the gas.

If you were inside the city limits looking at the highway, you would see everything at the city speeds. But would you notice a difference if you were to go to the edge of the city and watch the open-road section of the highway? From my own observations, no. Out on the open road, I can see quite a bit of difference. Would we get different results about the speed of light if we viewed it from outside the heliosphere? Would interstellar space be considered the open road where light can travel even faster?

I wonder if all solar systems have heliosphere-type bubbles around them. We know the heliosphere protects us from all the powerful radiation emanating from the universe. With that in mind, it's probably safe to say that all solar systems have them. I guess you can compare it to a gigantic filter that keeps the harmful radiation from reaching us on planet

Earth. Anything that passes through this filter would slow down. Our observations of starlight are of light that has passed through the filter of the heliosphere. I theorize the heliosphere also slows the speed of light to the speeds we currently observe.

The only way to be sure one way or another is to measure the speed of light from outside our own solar system. It seems the only thing holding us back is our current level of technology. I'm not complaining, because I enjoy what we do have. Honestly, I wouldn't mind being born one thousand years from now, or even five hundred years into the future. I expect by then we won't still be confined to Earth. I'm pretty sure one of the first things that will be measured in interstellar space is the speed of light.

If it happens that I'm right and the speed of light does increase in interstellar space, that means the universe itself could be even bigger and older than previously thought. That would be a pretty big discovery to say the least. The images from the James Webb Telescope make me wonder. The oldest galaxies, it seems, were present around two hundred million years after the big bang at a time when researchers were thinking star formation was just beginning. How can that be unless our calculations are wrong?

Now I'm starting to wonder about the magnetic tunnel that passes through our solar system. Not only are we connected to it, but it also connects to everything. Could this be a part of that interstellar highway I mentioned? At least it's proof everything is connected in some way, shape, or form. I would suggest looking into the research of Jennifer West, a researcher for the Dunlap Institute for Astronomy & Astrophysics.

There is the possibility that there are things that regularly travel faster than light. It could very well be that due to the speeds at which they travel, they exist outside our detection spectrum. If an object is traveling faster than light, that would be beyond our visual spectrum because it's moving faster than our brains can register it. So in effect, we would not see it.

What if the speed of light is just the threshold of our own mind? What if we can see only things that move slower than light? That could mean there is an undiscovered world that we can't access because it's literally moving too fast for us. Maybe that is why researchers are having such a hard time defining what dark matter is. Maybe whatever dark matter is could hypothetically be traveling faster than light. That could explain why we cannot see dark matter. We know some other force is involved, but what?

Could we ever build instruments that can pick up on speeds beyond anything that the human mind can identify? That is a question that cannot be answered logically. How do you look for something nobody knows exists? I mean, the only way to get a definitive answer would be to ask an alien, if they ever decided to come to Earth.

So as it stands at the moment, light is king when it comes to the speed of the universe. This can always change as we grow as a species. The real answers await us beyond the edge of our own solar system. One day humanity will reach beyond the light of our sun into the unknown. We could very well make this journey at speeds that exceed our current understanding of the universal speed limit. Remember, limits are meant to be pushed within boundaries. And light is another limit that will one day be pushed. If in fact the speed of light is the ultimate limit, space travel sure won't be as fun as *Star Trek* makes it seem.

Generational ships will be our only option. Honestly, the thought of that is rather boring. An entire lifetime spent in the confines of a ship without ever experiencing life the way we do. I imagine it would be psychologically difficult to spend generations on a ship to reach a new planet around a distant star. A lot of things could happen in the span of a thousand years, five hundred years even. What if the planet you were headed to became uninhabitable during your journey? What would you do? Try and find another planet? You might never find a home. Just another reason to continue my research into the speed of light.

CHAPTER 4

BEFORE THE BIG BANG

I'm pretty sure you know what black holes are by now, considering our universe is filled with them. Every galaxy has one at the center. The Milky Way has one in the Sagittarius constellation called Sagittarius A*. Ours is big, but small when compared to some of the other black holes out there.

The big bang theory basically explains how our universe came into being roughly fourteen billion years ago. That event gave rise to our solar system around four billion years ago. Without the big bang, nothing would exist, not even us humans. The big bang was the day our universe was born. Well, not exactly the day, because there were no days or time. In the absence of time, particles existed. Heat existed as well.

What if two separate things are connected, such as black holes and the big bang? You're probably wondering how in the heck black holes are connected to the big bang. I have a theory that might explain the connection. And if this theory is true, it would mean the rabbit hole is a deep one.

Black holes, monsters, behemoths, whatever you want to call them, are the most powerful things known in the universe. The one object that always wins in a tug-of-war for anything that wanders too close. Planets, stars, light, gas, and debris are just completely torn to shreds. There is no escaping the gravitational pull of a black hole. The gravitational pull is so strong it warps the very fabric of space itself. The immense power black holes have is amazing. They win against everything.

When you really think about it, black holes basically have all the building blocks of life. In some way, shape, or form, the information is all there. Imagine, over billions of years, everything that a black hole—with an appetite that never ceases—has consumed, condensed into a compact ball. That compact ball would have to have a tipping point. I would imagine that over time, black holes reach a point where they are stuffed. Even though they are stuffed, they continue eating.

If you or I tried to keep eating after our stomachs were full, we would get sick. Maybe some of us would even vomit. Our bodies, our stomachs, can hold only so much food. I myself am a decent-sized guy who likes to eat. Even I have had to back away from a meal or two in my day, no shame.

Applying that same logic to a black hole, I theorize that it can hold only so much. That amount is beyond anything we can use to measure it. We can tell you only how heavy a black hole is. We can't tell you if the stomach of a black hole is full. I wouldn't even know how to measure something like that. But I will come back to this in a minute.

The big bang happened roughly 13.7 billion years ago. That event is categorized as the birth of our universe and our very existence. A very chaotic event gave rise to such a beautiful, exotic universe. An event that expanded the universe so fast that if you blinked, you literally missed it. Everything expanded at speeds that make the actual speed of light appear as a school zone. There are no words to describe what seeing an event like that would be like.

On the larger scale of things, the big bang should not have happened. For everything to be so precise, at the right moment, to kick-start our universe is an anomaly to say the least. The chances of the big bang happening—let's just say the odds were against us. Yet here we are debating about it. It makes me wonder exactly how long things were mixing around before our universe exploded into existence. For all we know, it could have taken billions or trillions of years before the right combination of things created our universe. Of course, time did not exist in the way we know it.

The building blocks of life were here from the very beginning. How could they be present from the start? Unless our beginning was the ending somewhere else. What if our universe has a backstory beyond the bounds of our known reality?

What if the backstory begins with a black hole? Not just any black hole, but a special black hole beyond the reach of our very own universe. From another universe entirely separate from this one, sort of like a parent universe. A place where an ancient black hole ate for billions if not trillions of years. All of the building blocks of life, together in one place. Do you see where I'm going with this?

Let's just say this ancient black hole reached its limit. Its size was something we have never seen before in our own universe. It had a gravitational field so strong it was off the charts—we can't define it with current terminology. Possibly a black hole the size of an entire galaxy—a monster so big that it not only warped the fabric of space-time itself but also tore through it entirely. And this event is known to us as the big bang.

The big bang. A powerful force of energy that just exploded. There's really no other way to put it. Since it was only a one-time deal, we have only one example to work with. And because it happened such a long time ago, we have only scattered puzzle pieces everywhere, in a sense. We have pieced together some of that puzzle, but the majority of it remains a mystery.

I tend to think of the big bang event like a trash bag that became too heavy for everything inside. As the integrity of the bag becomes compromised, a stick of dynamite explodes in the middle. Next thing you know, all your garbage is all over the ground. A complete mess of things, I'll say. I believe that this is what happened with the creation of our own universe, just a lot more chaotic and on a massive scale.

All the information to create this universe had to come from somewhere. What if it so happens that our big bang was the result of an ancient black hole beyond our own space-time continuum? What would that universe be like? Would it be similar to our own universe? How old was it? Did intelligent life start there in a similar fashion to how it started on Earth? Was this universe compact? Or was it spread out farther than can ever be measured? How did time run? I can sit here and go on and on with questions.

In a sense, our big bang could be what is known as a hypothetical white hole, the total opposite of a black hole. A place where nothing would be able to enter. I theorize it would be like a firefighter's water hose turned on full blast. The closer you get, the more force is pushing against you, keeping you back. That could be the underlying reason our universe is expanding at the speeds it is.

Up to this point, scientists have not detected any white holes. Some scientists believe a white hole would appear as a black hole until it belched, or something like that. There could be a reason why we are not seeing white holes. A white hole could be just as rare as our own big bang event. Hence the possible connection between black holes and the big bang. What if we have already found that elusive white hole, our big bang?

That would mean that white holes are rare, unique phenomena that spawn entire universes. How awesome would that be? Finding the evidence to back this up is a different story. Since we haven't seen a white hole, there is no basis to go off. Since the big bang occurred in our distant past, our data is limited.

Now let's just say that white holes are the end result of ancient black holes from beyond our present universe. And those ancient black holes exist in a place and time that we know absolutely nothing about. So in theory, each parent universe could go on and on forever.

I wonder—when our big bang event happened, were other universes created at that same moment as well? In other words, the multiverse theory? I mean, it could be a possibility. Some scientists believe that they have found evidence of other universes. I theorize that we are a long way off from having solid proof of the multiverse.

Don't get me wrong—I believe in the theory of the multiverse, just a different version of the theory. My thoughts are that we all visit a part of this multiverse when we are fast asleep dreaming. I know there is no science to back up my claim, but there is also no evidence to disprove it either.

There have been a lot of times while I have been dreaming that things have appeared to be so real. Not only that, but I also return to the same place over and over again. Each time I visit this place, it's just like another day. While there, I remember every detail about the last time I was there—memories that I do not have, a familiarity with some people that feels as real as anything in this world. I interact with the same group of people. These people are people I have never met or even seen.

The city I'm in does not match anywhere I have been. It has aspects from several different cities mixed together. I would say it has the same level of technology as we have now. Somewhere, I suspect, this is a real place with a different version of me, somewhere in the multiverse. Maybe this place I have visited the most has a deep connection with me that I don't understand. What if that version of me visits me here and experiences my life with me? Like those times I have ideas that I know came from somewhere else, but they came from me, like when I have certain urges that I normally don't have.

I'm not crazy, if that's what you are wondering. I'm just open to possibilities that might be. Trying to find an explanation for things we can't explain at this time. No matter what you tend to believe, remember it all started with the big bang. So many different possibilities from one event. The more we search, the more we will find. One day we will know what was before the big bang. Perhaps everything we all thought we knew or theorized about the origin of the big bang could be wrong. Only time will tell.

Would we ever be able to enter into a hypothetical white hole and into the other side of the looking glass? Highly unlikely with our current level of technology and understanding. What if the hypothetical parent universe ended in the big freeze? What if an intelligent species created our big bang as a means of escape from a dying universe? Would humanity have to one day make that journey if we could?

CHAPTER 5

EVOLUTION OF DINOSAURS

Considering how far humans have come in a relatively short amount of time, why not dinosaurs? They had millions of years on this earth compared to us, so why couldn't they? I know it might sound crazy. But what if a species, or even several different types of dinosaurs, evolved into intelligent beings? What if they evolved long before the asteroid that hit sixty-five million years ago? They might have escaped to another solar system in an entirely different part of the universe for all we know.

Quite frankly, I do not believe we are the only intelligent life to have called planet Earth home. Our planet has been around for billions of years. That's a long time for us on Earth. In terms of space as a whole, our planet and our solar system are just babies at 4.5 billion years old. Our Milky Way galaxy is 13.6 billion years old, so we are on the young side. According to the cosmic background radiation, our universe is 13.7 billion years old. So basically, our galaxy formed around one hundred million years after the big bang event occurred. I just wanted to include the time reference in terms of the big picture.

The age of the dinosaurs lasted roughly 165 million years. That is way more than enough time for at least one dinosaur species to evolve into intelligent life. The first dinosaurs appeared 245 million years ago in what we know as the Triassic Period, which existed from 252 million years ago to 201 million years ago, followed by the Jurassic Period from about 201 to 145 million years ago, followed by the Cretaceous Period between 145 and 66 million years ago. These are the three periods of the Mesozoic Era. Plenty of time for intelligent evolution to occur, I theorize.

Honestly, they would not look like the modern depictions of dinosaurs we have. Do modern humans resemble our primate ancestors? Not really. And neither would an evolved dinosaur species. This is my theory, of course. Look at how much further we evolved beyond our primate ancestors. Look at the modern primates who more or less are closer to our shared ancestor.

Something big happened in our past that caused a mutation in our genes that gave rise to our intelligence. We have had only a tiny fraction of time on this earth compared to the time of the dinosaurs, and look how far we have come. Imagine the advances we could make after millions of years.

All it takes is one species to evolve, just like we did. If we view the dinosaur age in the same light as ours, it is possible. So where is the evidence? Depending on how you look at things, there are some out-of-place artifacts that are older than any known civilization on earth. A few of my favorite examples are the pre-Columbian Quimbaya airplanes, the Kingoodie hammer, and the Antikythera mechanism. The list goes on, and I won't get into that too much right now because there is a lot to take in. I do recommend looking into this subject in the future.

Also, there have been footprints fossilized right along with dinosaur tracks. There have even been a few fossilized soles of shoes found. Those items must have been present a long time ago to have become fossilized.

Combined with the other out-of-place artifacts, this fossil evidence suggests some form of intelligence was present on Earth millions of years ago. Whether it be humans or another humanoid life-form, until we have a fossil, we cannot say for sure either way.

The whole dinosaur-evolution theory reminds me of a *Star Trek Voyager* episode from season three titled "Distant Origins." I don't want to give too many details if you have not watched it yet. In the Delta Quadrant, there is a reptilian species called the Voth. Basically, a couple of Voth scientists find the skeletal remains of a crew member from *Voyager*. They become fascinated by the similarities of their genome compared to ours. Those same two scientists have a theory that their species came from somewhere else and colonized the planet they view as their home world. Again, I don't want to spoil the episode for you.

After watching that episode years ago, I began to wonder if something like that could have happened for real. A genetic cousin that had also evolved into an intelligent species on planet Earth. How cool would that be, to find out that we were not the only smart ones here?

What if Earth is just a nest? The planet where we evolve into higher intelligence before leaving for a new permanent home, kind of like a bird in the nest? When the time comes, the bird has to leave the nest and learn to fly on its own. What if we are the birds, so to speak? That would indicate this is only the beginning. Maybe leaving Earth would be considered leaving the nest. Learning how to travel faster than light could be humans learning to fly.

We struggle over resources in our current day and age. As the effects of climate change worsen, humanity will need to find a new planet to live on that will sustain the human race for thousands and thousands of years, if not millions. It also makes me wonder if climate change is a repeating theme. Hypothetically, what if the climate started becoming unbearable for a humanoid dinosaur species, and they were forced to leave Earth?

The asteroid sixty-five million years ago that wiped out the remnants of the dinosaurs was just a planetary reset, so to speak—the final nail in the coffin. Kill off the old to make way for the rise of another species that eventually evolved into us modern humans.

All the fossil fuels we currently use, such as coal, oil, and natural gas, were formed from organic material, living things on Earth, millions of years ago. What if the same fossil fuels we use today were tapped millions of years ago by an intelligent dinosaur species? Considering how long fossil fuels take to form, it could very well be a possibility. The earth has to replenish its resources somehow.

In theory, any evidence to prove that dinosaurs did evolve into intelligent, humanoid-like species is being used to fuel our modern-day world. If the earth exists far off into the future, then one day we might be the fuel to power a new species throughout their early modernization. In a sense, it would be considered recycling, just on a massive scale.

I'll step a little further outside the box to show another angle. I know a lot of people do not consider the alien abduction stories as real events that happened. In some instances, I believe what the abductees have said. There has always been talk about a reptilian species that has visited Earth. Some people have come forward with abduction stories linked to this reptilian species. These beings are described as large, humanoid-type lizards.

The name *dinosaur* itself comes from the Greek words for "monstrous lizard." I thought to myself, Could these be dinosaurs these people are describing? Could they be the descendants of the same dinosaurs that roamed our earth millions of years ago? Could this mean they did not all die out sixty-five million years ago when the asteroid struck earth? Are they now returning to see if their ancestral home world still exists?

I remember listening to some abduction stories in which a few separate people from around the world, who have never met, shared a common theme in their stories about these so-called reptilian aliens. The thing that really stuck out to me was that these abductees claimed that these reptilians told them that they are the descendants of what we call dinosaurs. There is no way to verify the claims of these people at all. So could they be telling the truth? Quite possibly.

The entire picture of ancient dinosaurs evolving into an intelligent, humanoid-type species is a big pill to swallow. When you think about it, it makes perfect sense. I know you have probably been wondering how this is possible when we clearly have plenty of fossilized dinosaur bones on display. I theorize that at least three distinct types of dinosaurs evolved into intelligent humanoid-like species. Possibly one from each of the three distinct time periods the dinosaurs existed in. The Triassic, Jurassic, and Cretaceous periods could have given rise to species that would evolve into more. Or the intelligent humanoid-like species could have a shared ancestor linking them to dinosaurs in the same way humans have a shared ancestor linking us to primates.

The fact is that most dinosaurs did not evolve. Humans do not regard any types of primates as equals. We went on and evolved an intelligence that is not seen among other life-forms on Earth. If we had the capabilities to leave Earth and there was an extinction-level threat upon us, I am pretty sure we would not be worrying about saving all the primates or other animals. We would be worrying about taking every person on Earth for the purpose of self-preservation of the human race. Millions of years down the road another, if another life-form found the bones left behind here on Earth, it might take human bones and bones from primates, given the genetic similarities, and conclude we are all just different types of monkeys and leave it at that. As a human, I know that is not the case with our species.

But in reality, that is exactly what we did with dinosaurs. We found bones and pieced them together. Have we found all the bones that are hidden around the world? Not even close. I believe we are just scratching the surface of what we know about the ancient world. So it really needs to be looked at through a different lens. Scientifically speaking, this is a real possibility. If we use our own human evolutionary path as a guide in reference to dinosaurs, there is a very good chance a species of dinosaurs evolved in a similar fashion as we did. Since we have only ourselves as a definitive model of what intelligent life is, we have to be open-minded as to how other life can evolve.

As we are currently learning through constant research, microscopic life-forms can live and thrive in environments that we previously thought nothing could survive in. Those same microscopic life-forms will one day evolve into many new species that will populate a future earth after the absence of humans. Maybe the way these life-forms are developing in what we would consider hostile conditions could be an indication of what the earth will be like millions of years from now. Whatever different species does arise would be completely alien to even our wildest imagination. Just going off the conditions they are evolving in, I think they will be able to withstand environments no human alive ever could.

What if ancient dinosaurs, the ones that evolved, saw microscopic life-forms in things they deemed inhospitable for them? Those same environments could have been conditioning all the life on this planet as we know it for millions of years so we could live on the earth we inhabit today. It would almost sound as if humanity was planned out long ago by our planet to exist. The exact same thing is happening now for a planned, future species that will rise long after humanity has spread throughout the galaxy, forgetting that the human race began on Earth in its—or should I say in *our*—distant past. The tales of Earth would be nothing more than myths and legends, remembered for entertainment purposes instead of historical facts. The same way most people view our myths and legends of the ancient world.

CHAPTER 6

ANCIENT HUMANS

The story of humanity is a complicated one because we really do not know the story of our origin. We have mythologies, religion, oral history, and the fossil record. Each of these aspects is separate from the others. If you take into account the different mythologies alone, you can see there are many different avenues one can take. But each avenue is a different perspective on the greater tapestry of our history. Add all the religious doctrines, and you have a complete mess. But what if the truth is actually somewhere in the middle of it all?

When we examine the fossil record of humans, it gives us a timeline of roughly six million years, starting with the primate group *Ardipithecus,* while our own Christian Bible puts the age of humans at around six thousand years old. There is a huge discrepancy between the two figures no matter how you look at it. The similarities are what we have to look at if we are to understand the story of humanity. As you can see, the number six is relevant to both, but why? Could it possibly be that the number six is central to the entire theme and not the zeros behind it? What if we are the sixth creation of life on earth? The sixth universe, perhaps?

Six million years ago is when our fossil record started. Of course, there were not any modern humans walking around. It took millions of years of evolution to bring us to what we are today, *Homo sapiens*. The *Homo* genus to which we belong started to distinguish itself from our primate ancestors more than two million years ago.

The oldest *Homo sapiens* fossils trace back to around 315,000 years ago. That is more than 100,000 years earlier than previously believed. These fossils were found in Morocco, at the Jebel Irhoud excavation site, which was discovered in 1960. Another site has remains that date to approximately 233,000 years ago in the eastern African country of Ethiopia, in the Lower Omo Valley region. A 260,000-year-old skull was found in Florisbad, South Africa. It appears that *Homo sapiens* was widespread across the entire African continent much earlier than was previously believed.

It changes the narrative completely. It is safe to say that there is a lot more to our history than we are aware of with all the new discoveries that are being made all the time. I imagine that within the next ten years the fossil record for *Homo sapiens* will be over half a million years old. With the help of artificial intelligence, we are going to learn a lot about how humans have evolved.

What if these fossils are just the tip of the iceberg? What if *Homo sapiens* is actually a lot older? Ancient humans were using fire at least a million years ago. Not just for tool making, but also for cooking food. Most people will dismiss this attainment as insignificant. In actuality, this is very big! Only humans cook food! I have never seen or heard of any primates, or any other animals, cooking food, let alone using fire. How would ancient humans know to cook food unless they had a high level of intelligence or they were taught. But taught by whom?

What if humanity was already in an evolved state by then? What if ancient humans were further along than we are now technology-wise?

Maybe some event in the distant past changed the course of humanity forever. There are several different possibilities that could have happened, such as a technological malfunction, a natural disaster, war among ourselves, war with an alien species, or an unknown disease. But there is clear evidence of humanity existing in our distant past.

Let's say that millions or perhaps billions of years ago, humans reached a level of technology that is beyond our wildest dreams. That time would have a different landscape from the world we live in. I would imagine a world with unlimited clean energy, no disease, no war. Human colonies in our solar system and possibly beyond. People just living the best life possible without a worry in the world. Machines would do the physically demanding jobs. Artificial intelligence and automation would control the working aspects of life.

It would have been an almost perfect world, a world of living in the moment as if tomorrow did not exist. A life of pleasure due to the advanced technology. Over time those same pleasures would be taken for granted. Expecting that things would never change would lead to a comfort level that could lead to carelessness, which always leads to big mistakes.

A few thousand years of machines running everything without needing to be fixed would make people lazy. Eventually people would forget how to make the machines run and the technology behind them. Over time they would not see the need to learn to continue making things work, things that never break. They might have figured the computer would always have the answer. One day, after thousands or even millions of years, something could have malfunctioned. I do not mean a small problem; I am talking about the big one.

A possible energy source they were using could have exploded, taking most things with it. The few people that could have survived an event like that would have been the people that lived what we would call a rural lifestyle with limited technology. While the world around them ended, they continued and thrived because they chose to live a simple life. Or perhaps the survivors learned to adapt because the only other option was death.

It would be like if, for instance, some event happened that destroyed our technological grid: the Amish people would continue to thrive because they chose to live a life without all the modern comforts we enjoy now. Homesteaders and mountain men as well. They would experience no change at all in their quality of life. Now for the rest of us, it would not be that easy. I imagine with the way people are dependent on their phones, a lot of people would have a hard time adjusting to life that was not at their fingertips. It would be a serious reality check for some people. The modern world would be reduced to just a memory.

What if it was not a technological malfunction at all? Perhaps the asteroid that killed the dinosaurs wiped out most of humanity as well. If this were true, then that would mean that the lineage of humanity stretches back millions of years into antiquity. A discovery of this magnitude would change everything that we know about the past. The dinosaur-killing asteroid basically destroyed the earth.

Over 75 percent of life went extinct during that time. Imagine the amount of destruction that occurred. When I look outside and see all the nature around me, it's hard to picture that what I'm seeing is only 25 percent of what survived. The earth was truly an alien world compared to the earth we know now.

When you think about it, that event would have destroyed not only most of life but also everything in general. With nobody around to maintain the infrastructure if any was left, everything would have stopped working and would eventually have crumbled away. Tectonic plates shifting would have resulted in a lot of things being buried. Nature would have reclaimed everything else that wasn't buried. There would have been no trace left of anything after a while. Almost as if nobody existed at all. A complete reset of the earth.

But what if those few that survived were just a different branch in the tree of humanity? A more evolved *Homo sapiens*, perhaps? And the few that did survive mated with our genetic cousins so as to not allow the extinction of the human race. We would be the product of those efforts. Not exactly like them, but a very close descendant—a mixture of them and a distant *Homo* species. We do share 98.8 percent of our DNA with chimpanzees. That 1.2 percent is what separates us from them.

Maybe that could explain the origin of our intelligence. This is a mystery that science cannot explain at the moment. What we do know is that our brains grew in size over time. As a result, we became smarter—at least that is the main theory. It is almost like our intelligence occurred overnight. But how, and why? Or maybe our intelligent ancestors did not mate with the other branches, and we are exactly as they were. At this point we do not know. We can only speculate based on the fragments of information we currently have. The more understanding we have, the more questions come from it.

Or what if things had a much darker tone, such as war? With the current conflict in Ukraine and the threat of nuclear war, this is a real possibility. One country has been building the most destructive nuclear weapons known to man, threatening the very existence of humanity itself if it doesn't get its way. This is a very dangerous scenario that I fear could be playing out again. People in positions of power dooming the rest of us out of selfish ambitions. Some people are willing to end the world if they cannot rule it. I believe this is not the first time humanity has faced a devastating war.

In fact, there is evidence around the world of nuclear war millions of years ago. More specifically, trinitite or Alamogordo glass is the glassy residue left after nuclear explosions. The only way this can form is through the extreme heat that can be produced only in a nuclear-type blast. Any chance of this happening naturally? No. And there are various sites around the world that have trinitite dating back to before our civilization started.

It would not be far-fetched to assume something like this did happen. In fact, it is a very real possibility. Some religions describe ancient weapons with immense, destructive capabilities. Some stories even mention the use of some of these weapons. From a scientific standpoint, more examination is needed. Maybe these stories are more than just stories. Maybe these stories are telling us what actually happened in the distant past. Some ancient pictures depict certain items that look very similar to modern pictures of certain nuclear aspects.

However, with our current level of technology, we have harnessed similar weapons. Look at the nuclear weapons that some countries have developed. Look at the destruction we unleashed on Japan during World War II with two bombs. Our nuclear weapons nowadays can do so much more damage in a much smaller package. These weapons were designed to kill a lot of other humans, not extraterrestrials from another planet. And as our technology progresses, these weapons we continue to develop will become that much more destructive.

That same kind of thinking could have brought destruction to a previous civilization. The ones who hold the biggest weapons are usually the ones with the most power. Nuclear weapons are seen as status symbols of powerful nations. Maybe that kind of thinking is not just human. It could very well be that the acquisition of power by means of force is a concept shared across the universe by intelligent species. Then again, it might not be.

Well, let's just say it is a shared concept among some species of the universe. Then conquest and war with other star systems would be a natural way of life. Could an ancient war with a species not from Earth have happened? The simple answer is yes, a war that destroyed civilization and reset humanity to the beginning. A war that left its scars across our solar system and the legends and mythologies around the world.

What if humanity had its origins billions of years ago? A human species that left Earth and spread out across the galaxy? Sounds almost like *Star Wars*. Except it did not happen in a galaxy far, far away. It happened right here in our very own Milky Way galaxy. What if ancient humans did colonize the Milky Way, only to find a much superior species that fought them back to their home world, destroying every colony and every bit of technology humans had and sending humanity back to square one? The only humans that remained were left alive on Earth as a reminder of what happened.

So many clues point to humanity having existed well before now. *The Cosmic War* by Joseph P. Farrell is a great read. In fact, it opened my mind up to other possibilities.

I'm going to start with the exploded planet hypothesis. When you look at the planets in our solar system, you can easily see a pattern. Mercury, Venus, Earth, Mars, and then a big gap before you reach Jupiter and the outer planets. In the middle of that gap is where the asteroid belt lies, in the same place a planet should be. How do we know that another planet would not/should not be there?

The Titius-Bode law is a formulaic prediction of spacing between planets in any given solar system. The formula suggests that, extending outward, each planet should be approximately twice as far from the sun as the one before.

This idea was first announced to the world by German astronomer Johann Daniel Titius in the year 1766. Johann Elert Bode, also from Germany, made the theory popular in 1772. Hence the name Titius-Bode law, known more commonly as Bode's law. I'm not saying Bode's law is perfect. But when it comes to our own solar system, it does raise some questions.

The asteroid belt is something of a big mystery to me. I find it hard to believe that the asteroid belt is remnants from a protoplanetary disk that never formed. All the other planets in our solar system formed, but why not this one? To me, it looks like the debris from a previous planet. Are these all the pieces of that planet? No! I suspect that some chunks of that planet became moons of Jupiter. Maybe some chunks even ended up in Jupiter's atmosphere. I predict that one day this connection will be made.

Let's say that in our ancient past, our solar system was much more than it is now. This hypothetical planet was sitting between Mars and Jupiter, possibly back when Mars could have supported life. Ancient humans might have inhabited all three plants, Earth, Mars, and this hypothetical planet. What if an ancient interplanetary war happened that resulted in the complete destruction of this hypothetical planet? This really could explain how Mars lost its atmosphere. It was blasted away, killing everything that once thrived.

We do know that ancient Mars had an atmosphere and water. When you look at some of the pictures of Mars, you can see evidence of an ancient, extinct civilization. Architecture, shapes, patterns, pyramids, the face, among other things. Almost as if it was once a thriving world but is now completely destroyed—a graveyard frozen in time. Some areas look like they were bombed repeatedly.

Mars, the god of war in Roman mythology, has a gash on his thigh. Mars the planet also has a gash that is named Valles Marineris. In fact, it is the largest canyon in our entire solar system, easily dwarfing the Grand Canyon. Maybe there is a connection between the two.

CHAPTER 7

THE ABSENCE OF EXTRATERRESTRIAL COMMUNICATION

The big question is why we have not picked up on any signs of intelligent life in space. No radio signals, just nothing. We have all these telescopes searching day and night for some sign that we are not alone in the universe. Unfortunately, we are still left in the dark regarding extraterrestrial communication.

Have you heard about the Wow! signal that Ohio State University's Big Ear radio telescope picked up on August 15, 1977? If not, don't worry; I will give you the basics. However, I do recommend researching this on your own.

In reference to the Wow! signal, we do know that it came from the direction of the constellation Sagittarius. The Wow! signal was a radio signal detected on August 15, 1977, by Ohio State University's Big Ear radio telescope. The signal came from the direction of the Sagittarius constellation and appeared to be extraterrestrial in origin. It lasted for seventy-two seconds and has never been detected again. There are a lot of theories about the origin of the signal. I am not going to echo what has already been said. You can say the same thing only so many times. Instead, I would like to approach this from a different angle and possibly breathe new fire into this.

Out of the vastness of the universe, only the Wow! signal has been found. How can that be? I theorize that it is impossible that there is not any extraterrestrial communication going on around us and throughout the universe in general. There are galaxies upon galaxies, and the stars are too numerous for a person to count in ten lifetimes.

Most of these stars have planets. Some are in what is referred to as the Goldilocks zone—a place around a star where the climate would be comparable to that of Earth. Within this zone, liquid water is thought to exist. Finding planets within this Goldilocks region around stars is very important in our search for life beyond Earth. In reality, we will not know for sure until we venture out into space and see for ourselves. The only thing we can do is theorize about what is out there based on the data available for our level of technology.

What if life can evolve in different conditions from those on Earth? What if poisonous gasses are what extraterrestrials breathe? Maybe liquid methane would be like water to them. Maybe they live on moons of planets that are outside the habitable zone. Or even inside planets or moons.

You can look at pictures from the 1800s and early 1900s, and nothing is the same anymore. You can literally stand in the exact same spot an old photo was taken; you would not recognize the area today. Do you think a television from the 1960s could pick up our television channels in 2023? No, it could not. I will even go back to 2005. Those televisions from 2005 cannot pick up our modern stations, and that was less than twenty years ago. So not that long ago, but different technologies are being used.

Do you see where I am going with this? An analog television cannot pick up digital television channels. Same stations, same planet, but different results depending on the technology you use. The only way to pick up a digital broadcast on an analog television is with a digital converter box. Even with the converter box, it isn't the best signal.

Our technology cannot pick up on extraterrestrial communication because we are too primitive. We are looking at things from our limited understanding. I am willing to bet there is a whole other side of physics that we don't know exists yet. Our technological knowledge is in its infancy. We are not advanced enough to even begin to understand the level of technology being used by extraterrestrials.

Imagine the advances a hundred years from now. A thousand? It would be like stepping into the wildest science fiction adventure you could imagine. As a teenager in the '90s, I would have never imagined some of the things we have now, such as the pocket-sized computers we call cell phones, wireless charging, Wi-Fi, and video games that are becoming more realistic. I mean, I could go on about things we have now, that would have blown my mind back in the '90s.

We are advancing at a rapid pace. My children are going to experience things I watch in science fiction movies all the time. Hopefully, sometime in the relative near future, we will be advanced enough to eavesdrop on other civilizations. Some people believe we are alone. I myself do not.

Let's say we are not alone, and the universe is filled with different types of intelligent species. Imagine all the possibilities that life could have risen from. Some species would be biologically different from humans. Some would be similar to us in appearance.

If a species were to evolve into a hypothetical type 2 or type 3 civilization and had the technology to branch out among the stars, they would have to have an elaborate communication system that we cannot even begin to comprehend. A system that is invisible to our modern technology. A system that is based on a physics that we know absolutely nothing about because we do not know it exists yet. I am guessing that in a thousand years, our knowledge of physics will be at that point to understand how a system like that could even work. We would even know how to travel the stars within a lifetime.

Maybe extraterrestrial transmissions do not travel through space like ours would. What if there is a layer of space or a string that connects everything under the surface of what we see? And maybe this layer of subspace acts very differently than regular space. What if this layer of subspace allows for instantaneous transmission signals? Or they might use the same technology they use for interstellar travel to send signals back and forth. All the possibilities, but nothing concrete as of yet.

One thing we do know is that there is a vast distance between stars in our own galaxy, as well as in other galaxies. Comprehending that distance is difficult; it is really off the scale compared to how far we can travel. Imagine going to another planet. Just think about the distance to another solar system: it's enormous. Just trying to send a message from planet to planet would be a big deal to us because we have not figured it out yet. I am not talking about Mars or somewhere close. I am talking about a planet in a completely different solar system!

It makes me wonder if it could just be that our timing is off. Consider the distance it takes for the light from a distant star to reach us. The starlight we see every night is a look into the past. We do not see how the stars currently look because the light from those distant objects had to travel all that distance to reach us. For example, the Andromeda galaxy is 2.5 million light-years from us. When we see the light from that galaxy, we are seeing it as it was 2.5 million years ago because that is how long the light took to reach us. It is pretty amazing when you really think about it. What if some of those stars do not even exist anymore? This is a very real possibility.

What if an intelligent species already rose up and conquered their home galaxy, or at least a local region of their galaxy, and fell into extinction over the course of, let's say, fifty thousand years? What if their transmissions to Earth have not even reached us yet? And by the time we receive them and can respond, we are responding to nothing?

This one puts a chill in my spine. What if a superior species came along and wiped them out, and they sent out a message warning other worlds not to suffer the same fate as they did? By the time the message comes, it will already be too late—we will have been wiped out as well. I do suspect this is not the case.

What if we instead turn our attention to stellar nurseries? Of course those stars would not look the same as we see them now. But it might be a good place in time to pick up on our galactic cousins. This, of course, is just my opinion. We might catch blips from civilizations that are technologically on par with us. And just maybe, we will be able to pick up some kind of extraterrestrial signal that would be the smoking gun that we are not alone in the universe.

Will we figure out the biggest question of all—are we alone? Yes, we will! When that day comes, it will change a lot of things. It will truly be a giant leap for mankind. To finally know that we are not alone would change the course of history. To know that we are only one part of the bigger puzzle would truly deepen the mystery of where we came from, as well as where they came from. It would mean that we are connected to something much bigger than humanity itself.

CHAPTER 8

TIME TRAVEL

One of the wildest concepts of science fiction has been time travel. For many years it remained in the realm of science fiction. Over the years, as our understanding of physics has broadened, things that were deemed impossible have finally made it to the realm of possibility. One such subject happens to be time travel. We might not have the resources or the know-how to build a time machine. Theoretically, science says it is possible.

When we normally think about time travel, most of us would probably think of *Back to the Future* or some other movie we have seen in our lifetimes. Imagination and special effects gave us a pretty good show. Truthfully, the thought of time travel was with us much earlier thanks to the mind of H. G. Wells. He wrote the book *The Time Machine* back in 1895, a great book that gave rise to a popular genre more than one hundred years later. If you have not read it, I strongly recommend that you do.

When you think of the different aspects of time travel, it's a little more than complicated. For one thing, what is time? The simple answer is the duration in which all things happen. Not much of an explanation, but still, what is time? Is time an illusion that we all share? Is time linear or stacked, like I explained in chapter 2? I suspect it is much more complicated than that.

According to Albert Einstein, you need to describe where you are not only in three-dimensional space—length, width, and height—but also in time. Time is the fourth dimension. So to know where you are, you have to know what time it is as well.

What if space is just a visual representation of time? Space and time are intertwined no matter how you look at them. If you travel in time, you are also traveling in space, and vice versa. Since they are so intertwined with each other, it could really be two sides of the same coin. If you can manipulate one, you can manipulate the other. The more we learn about quantum mechanics, the more I suspect this will one day be the accepted answer.

Let's say you are in space, traveling at nearly the speed of light. For you, time will have slowed down quite a bit in comparison to a person who is on Earth. This is called time dilation. Time is not constant. Time is experienced differently by everyone. Since time is relative to the user, we should be able to manipulate time. The question is how?

A simple twenty-year round-trip journey in space would be a one-way trip to the future. Everyone you ever knew on Earth will have long passed away, while you will have aged only those twenty years while you were gone. You would essentially be returning to a different world. More than one hundred years would have gone by. That would be scary, not knowing what you were returning to.

Another example of time dilation that has been observed and recorded has to do with a set of twins. Not just any twins, but astronauts. Scott Kelly and Mark Kelly are identical twin brothers. Both brothers are former NASA astronauts. Scott Kelly spent a whopping 520 days orbiting Earth on the International Space Station. The International Space

Station orbits the earth at 17,500 miles per hour (28,160 kilometers per hour). That is very fast, no matter how you look at it. Most of us will never get to experience what going that fast feels like.

Out of the two brothers, Mark Kelly is older, having been born six minutes earlier. After Scott spent 520 days in orbit of Earth, Mark is now older than his twin brother by six minutes and five milliseconds. I know it does not seem like a significant difference, but it proves Albert Einstein right with his theory about time dilation.

So theoretically, you could park a spaceship next to a black hole and travel into the future, providing you did not get sucked in. Since the gravity of a black hole is so immense, time would be stretched. You would experience time much more slowly, and the time dilation effect would be much greater. After the time you preset had passed, you would be able to exit the gravity well at a much later time in the future. There you have it: time travel.

This is an example of time travel. A one-way trip into an uncertain future, to put it mildly. But I am guessing that this is not the time travel you were expecting. You were probably thinking about being able to visit any period in time—past, present, future. Well, that time might not be too far off in our future, thanks to Albert Einstein's theory of special relativity. Special relativity is an explanation of how speed affects mass, time, and space. The theory includes a way for the speed of light to define the relationship between energy and matter. Small amounts of mass (m) can be interchangeable with enormous amounts of energy (E), and (C) is the speed of light, as defined by the classic equation $E = mc^2$.

The theory of time travel by theoretical physicist Ronald Mallett is more or less centered on light. According to Ronald Mallett, "Just by our own observations of black holes we know that gravity affects time. And believe it or not, light can create gravity. So if light can create gravity, and if gravity can affect time, light should be able to affect time." It is a pretty interesting concept when you look at it scientifically. The only problem is the energy needed to put the theory to the test. The amount of energy needed is way beyond our current capabilities—the unfortunate roadblock to many scientific experiments.

A possible energy source is one that we have not tapped into, and that is the electrical fields of the earth. Spiders use the same electrical current to fly, not wind. Nikola Tesla tried to expose the world to this abundant energy source. The earth's electrical grid has a lot of power. All the thunderstorms around the world pack a lot of energy. A bolt of lightning is a lot of unharnessed energy. If we could figure out how to harness this energy source, we could do so much more scientifically.

But let's say the energy needed for time travel is not a problem. Getting the right coordinates is another obstacle to overcome. Remember that Earth, our solar system, our galaxy, and the universe itself are constantly moving. Earth will never be in the same place twice, and that is a fact. You have to figure out not only where, but when. Across four dimensions, I imagine. It would be the same basic principle for space travel, just on a much larger scale. One miscalculation could literally have you lost in time.

Another thing to consider is that you might be able to travel back only to when the time machine was first turned on. Anything before that might be beyond the boundaries of the time loop. So no going back to watch historical events, unfortunately. But what if this is wrong? What if there are no boundaries when it comes to time travel?

If there are no boundaries regarding time travel, then that would change everything. Every past event could be watched over and over again. Even the very beginning of our universe could be watched as it happened. That would be my preferred destination if I could travel through time. Seeing how everything came to be would be amazing. But that's just me personally. Of course there are a lot more places in time I would like to visit, but this would be the first stop.

But if there are no boundaries to hypothetical time travel, then what could go wrong? Possibly changing historical events to create a radical future where a lot of things are different? Just thinking about the possibility of changing things brings to mind a famous paradox known as the grandfather paradox.

One of the aspects of dealing with time travel would be the grandfather paradox. Imagine that you travel back in time and kill your grandfather so that your father was never born. Since your father was never born, you were never born, so how could you have traveled back in time to kill your grandfather? It cancels itself out. Scientists use this paradox as an example against time travel. When you really think about it, this paradox throws a wrench into time travel.

The question then turns to this: If we cannot change events, could we at least view them? Almost like a movie when you think about it. You can rewind to any part of a movie and watch that part over and over again without ever changing how the movie plays out because the movie was already made. What if that is the extent of time travel to the past? We can only watch an event, and that is it. Every event that has happened since that day in the past to the time when time travel becomes possible would be set in stone. Those events would have been set in stone, past tense.

But what if this is one of those instances where science is wrong? What if the grandfather paradox really does not exist? What if nature has its own backup system to ensure the things that are meant to happen will happen regardless of what you do? Let's say that you successfully travel back in time and kill your grandfather. Your grandmother ends up remarrying, and a version of your father is born. And this version of your dad marries your mom, and you still end up being born.

So in reality, you never killed your grandfather—you would have killed a complete stranger. You would probably still have memories of him. But he would have none of you. That would be creepy to say the least. What if the key to your existence has to do with your grandmother and not your grandfather?

Or what if your existence has absolutely nothing to do with your grandparents at all? I know this sounds crazy, but science has uncovered a clue that could make this possible. Scientists have recently found a surprise connection between doppelgängers. Not only do doppelgängers share physical features and personality traits, but they share DNA as well. So they are basically copies of each other with no shared family between them. This is very important because it would hypothetically cancel out the grandfather paradox.

If you were supposed to be born, you will be born, simple as that. I know that is difficult for some people to accept. But with the new understanding regarding doppelgängers, it really makes it seem that we are not that unique after all, in a general sense.

Now that we have covered time travel to the past, what about time travel to the future? How can you travel somewhere that is unknown, to a place that does not exist yet to us? For us in the present, the future changes every day. Each choice we make shapes that future outcome. If you remember my earlier chapter when I was talking about the book universe theory, you should already have your answer.

Let's say you can travel in time and change past events. From that point, the timeline would be split into two, the original and the new. So in reality your timeline would be unaffected. The timeline that would change as a result of what you did would be the new one. So in theory this would suggest a form of the multiverse.

How do we know that people aren't already traveling through time, changing things? Logically, nobody would know if changes were made in the timeline. How some events happened would just be puzzling when they didn't make sense.

At this point in time, I can say that nobody knows for sure. Will time travel one day be a reality? My thoughts are yes. Probably not the way we imagine it. Some form of time travel will one day exist. The outcome of those choices is a different story that is beyond me at this point.

CONCLUSION

In conclusion, I would say that we are alive at the greatest time in recorded history—the beginning of the technological age, which has drastically changed our outlook on life in general. We are living at a time when science fiction is becoming science fact.

Science is starting to unravel the mysteries of the cosmos and peer back into a time that stretches into the billions of years—a time of mystery.

As I close this book, I leave you with a new outlook on aspects of our very existence, the universe, time, and everything else I have touched on within these pages. A new approach to age-old questions. These theories that I have shared are meant to push the confines of the reality around you so you can try and understand new things.

The same thinking will always get the same results. Sometimes the wildest thoughts are closer to the truth. Thinking beyond the known and accepted facts is where discoveries are made. The people who dare to challenge what is known push humanity to the next step.

For humanity to one day leave Earth and travel among the stars, the people with the wildest ideas are going to make it happen. Not the people who play it safe and accept the limitations they are told we have.

Scientific achievements have always been about pushing the known boundaries.

The theories inside this book are meant to push those same boundaries. Some of these theories could very well be proven to be facts someday. Maybe one of you reading this book will be the one to prove one of these theories true.

I hope this book inspires something in you. A search for answers to questions that haven't been asked yet, or maybe an extension of what I have shared. One day this very book could be a reference to some new discovery in the future. Let's usher in this future together.

About the Author

Clayton Murph was born in June of 1983 in San Antonio, Texas. Raised in a military family, he spent much of his childhood moving around. His mother passed away in 1994, leading his father to settle down in Denver, Colorado. After marrying his wife in 2003, he became interested in researching the origin of the universe and deepening his knowledge of concepts by exploring alternative theories. He now lives in southwest Virginia with his wife and five children. When not working or spending time with his family, he reads and researches.